NATURAL DISASTER RESEARCH, PREDICTION AND MITIGATION

HAITI: EARTHQUAKE AND RESPONSE

NATURAL DISASTER RESEARCH, PREDICTION AND MITIGATION

Additional books in this series can be found on Nova's website under the Series tab.

Additional E-books in this series can be found on Nova's website under the E-books tab.

NATURAL DISASTER RESEARCH, PREDICTION AND MITIGATION

HAITI: EARTHQUAKE AND RESPONSE

RACHAEL A. DONLON
EDITOR

Nova Science Publishers, Inc.
New York

For permission to use material from this book please contact us:
Telephone 631-231-7269; Fax 631-231-8175
Web Site: http://www.novapublishers.com

Library of Congress Cataloging-in-Publication Data

Haiti : earthquake and response / editor, Rachael A. Donlon.
p. cm.
Includes index.
ISBN 978-1-61668-998-8 (hardcover)
1. Earthquake relief--Haiti. 2. Haiti Earthquake, Haiti, 2010. 3. Haiti--Economic conditions--21st century. 4. Haiti--Social conditions--21st century. 5. Haiti--Politics and government--21st century. I. Donlon, Rachael A.
HV600 2010. H2 H34 2011
363.34'95097294090512--dc23
2011033957

Published by Nova Science Publishers, Inc. † New York

CONTENTS

PREFACE

The largest earthquake ever recorded in Haiti devastated parts of the country, including the capital, on January 12, 2010. The quake, centered about 15 miles southwest of Port-au-Prince, was a magnitude of 7.0, followed by a series of severe aftershocks. It is estimated that 3 million people, approximately one third of the overall population, have been affected by the earthquake. The Government of Haiti is reporting an estimated 112,000 deaths and 194,000 injured. This book explores the Haitian earthquake and global response as immediate needs are met and the humanitarian relief operation continues. The government is struggling to restore the institutions needed for it to function, ensure political stability, and address long-term reconstruction and development planning.

Chapter 1- The largest earthquake ever recorded in Haiti devastated parts of the country, including the capital, on January 12, 2010. The quake, centered about 15 miles southwest of Port-au-Prince, had a magnitude of 7.0. A series of strong aftershocks have followed. Experts estimate the earthquake caused $8 to $14 billion in damage. Approximately 3 million people, roughly one-third of the overall population, have been affected by the earthquake with estimates ranging from 1.2 to 2 million people displaced. The government of Haiti is reporting an estimated 230,000 deaths and 300,600 injured. In the immediate aftermath of the earthquake, President Rene Préval described conditions in his country as "unimaginable," and appealed for international assistance. As the humanitarian relief operation continues, the government is struggling to restore the institutions needed for it to function, ensure political stability, and address long-term reconstruction and development planning. Prior to the earthquake, the international community was providing extensive development and humanitarian assistance to Haiti. With that assistance, the Haitian government had made significant progress in recent years in many areas of its development strategy. The destruction of Haiti's nascent infrastructure and other extensive damage caused by the earthquake will set back Haiti's development significantly. A post-disaster needs assessment estimated the total value of recovery and reconstruction needs to be $11.5 billion. The Haitian government presented an action plan for reconstruction and development at a global donors' conference held on March 31, 2010. Donors pledged over $5 billion for the first 18 months of Haiti's reconstruction. The United States pledged $1.2 billion. Extra-constitutional rule will begin after May 10, when most parliamentarians' terms expire; President Préval will probably rule by decree after that. There is no timetable for new parliamentary elections.

Chapter 2- On January 12, 2010, a magnitude 7.0 earthquake struck Haiti. As of January 25, 2010, the death toll was estimated to exceed 150,000. The earthquake and resulting aftershocks affected approximately 3 million people and caused significant damage to buildings and infrastructure. The earthquake has left an estimated 1 million Haitians homeless.

Chapter 3- The environmental, social, and political conditions in Haiti have long prompted congressional interest in U.S. policy on Haitian migrants, particularly those attempting to reach the United States by boat. While some observers assert that such arrivals by Haitians are a breach in border security, others maintain that these Haitians are asylum seekers following a decades old practice of Haitians coming by boat without legal immigration documents. Migrant interdiction and mandatory detention are key components of U.S. policy toward Haitian migrants, but human rights advocates express concern that Haitians are not afforded the same treatment as other asylum seekers.

Chapter 4- The 1994 Northridge (CA) earthquake caused as much as $26 billion (in 2005 dollars) in damage and was one of the costliest natural disasters to strike the United States. The Federal Emergency Management Agency has estimated that earthquakes cost the United States over $5 billion per year. A hypothetical scenario for a magnitude 7.8 earthquake in southern California estimated a possibility of 1,800 fatalities and over $200 billion in economic losses. The January 12, 2010, magnitude 7.0 earthquake that struck Haiti only 15 miles from Port-au-Prince, the capital city, has caused an estimated 230,000 fatalities and 300,000 injuries. On February 27, 2010, a magnitude 8.8 earthquake occurred offshore of Chile and has caused massive damage, although far fewer fatalities than the Haiti earthquake. The Cascadia Subduction Zone megathrust fault, which stretches from southern British Columbia to northern California, has the potential to generate a massive earthquake and tsunami, similar to the Chilean earthquake, that could threaten Oregon, Washington, and northern California.

Chapter 5- This is a statement of Chairman John F. Kerry, United States Senate Committee on Foreign Relations.

Chapter 6- This is a statement of Dick Lugar, United States Senator for Indiana.

Chapter 7- This is a statement of Dr. Paul Farmer to the United States Senate Committee on Foreign Relations.

Chapter 8- This is a statement of James Dobbins, before the Senate Foreign Relations Committee.

Chapter 9- This is a statement of M. Rony Francois, United States Senate, Committee on Foreign Relations.

Chapter 10- This is a statement of Robert Maguire, Trinity Washington University and The United States Institute of Peaces before the Subcommittee on International Development, Foreign Assistance, Economic Affairs and International Environmental Protection of the US Senate Committee on Foreign Relations.

Chapter 11- This is a statement of Mark L. Schneider, Senior Vice President, International Crisis Group to the Senate Foreign Relations Subcommittee hearing on International Development and Foreign Assistance, Economic Affairs, and International Environment Protection on Haiti Reconstruction: Smart Planning Moving Forward.

Chapter 12- This is a statement of Charles MacCormack, President and CEO, Save the Children, before the Senate Foreign Relations Subcommittee on International Development and Foreign Assistance, Economic Affairs and International Environmental Protection.

In: Haiti: Earthquake and Response
Editor: Rachael A. Donlon
ISBN: 978-1-61668-998-8

Chapter 1

HAITI EARTHQUAKE: CRISIS AND RESPONSE*

Rhoda Margesson and Maureen Taft-Morales

SUMMARY

The largest earthquake ever recorded in Haiti devastated parts of the country, including the capital, on January 12, 2010. The quake, centered about 15 miles southwest of Port-au-Prince, had a magnitude of 7.0. A series of strong aftershocks have followed. Experts estimate the earthquake caused $8 to $14 billion in damage. Approximately 3 million people, roughly one-third of the overall population, have been affected by the earthquake with estimates ranging from 1.2 to 2 million people displaced. The government of Haiti is reporting an estimated 230,000 deaths and 300,600 injured. In the immediate aftermath of the earthquake, President Rene Préval described conditions in his country as "unimaginable," and appealed for international assistance. As the humanitarian relief operation continues, the government is struggling to restore the institutions needed for it to function, ensure political stability, and address long-term reconstruction and development planning. Prior to the earthquake, the international community was providing extensive development and humanitarian assistance to Haiti. With that assistance, the Haitian government had made significant progress in recent years in many areas of its development strategy. The destruction of Haiti's nascent infrastructure and other extensive damage caused by the earthquake will set back Haiti's development significantly. A post-disaster needs assessment estimated the total value of recovery and reconstruction needs to be $11.5 billion. The Haitian government presented an action plan for reconstruction and development at a global donors' conference held on March 31, 2010. Donors pledged over $5 billion for the first 18 months of Haiti's reconstruction. The United States pledged $1.2 billion. Extra-constitutional rule will begin after May 10, when most parliamentarians' terms expire; President Préval will probably rule by decree after that. There is no timetable for new parliamentary elections.

The sheer scale of the relief effort in Haiti has brought together tremendous capacity and willingness to help. As the rainy and hurricane seasons begin, the massive humanitarian relief operation underway in Haiti is focused on providing waterproof

* This is an edited, reformatted and augmented version of a Congressional Research Services publication, dated May 6, 2010.

emergency shelter, improving sanitation and meeting the basic needs of the displaced and other vulnerable Haitians. The relief effort is expected to last for many months. On January 12, 2010, President Barack Obama assembled heads of U.S. agencies to begin working immediately on a coordinated response to the disaster, with the U.S. Agency for International Development through the Office of Foreign Disaster Assistance as the lead agency. On January 14, the Administration announced $100 million in humanitarian assistance to Haiti to meet the immediate needs on the ground. As of April 30, 2010, total humanitarian funding provided to Haiti for the earthquake had reached over $1 billion. In the FY2010 supplemental request, the Administration is seeking a total of $2.8 billion for Haiti. Of that, $1.5 billion is for relief and disaster assistance, which would reimburse U.S. government agencies for services provided and for funds already obligated for ongoing relief activities. The request for recovery and reconstruction is $1.1 billion. This is primarily for new activities, focused on urgent infrastructure repairs, especially in the energy and agricultural sectors; critical health care; governance; and security.

The Department of Homeland Security has temporarily halted the deportation of Haitians and granted Temporary Protected Status for 18 months to Haitian nationals in the United States as of January 12, 2010. Congressional concerns include budget priorities and oversight, burden-sharing, immigration, tax incentives for charitable donations, trade preferences for Haiti, and helping constituents with pending adoptions and other issues. Several congressional committees have held hearings on Haiti. The focus of this report is on the immediate crisis in Haiti as a result of the earthquake, the U.S. and international response to date, and long-term implications of the earthquake. Related legislation includes P.L. 111-117, P.L. 111-126, P.L. 111-158,H.R. 144, H.R. 264, H.R. 417. H.R. 1567, H.R. 3077, H.R. 4206, H.R. 4577, H.R. 4616, H.R. 4952, H.R. 4961, H.R. 5006, H.R. 5160, H.R. 5171, S. 2949, S. 2961, and S. 2978, S. 2998, S. 3202, and S. 3275.

RECENT DEVELOPMENTS

- A major global donors' conference was held March 31, 2010, in New York. Donors pledged over $5 billion for the first 18 months of Haiti's reconstruction. The United States pledged $1.2 billion.
- The Haitian government presented an action plan for reconstruction and development, and a Post Disaster Needs Assessment that estimated the total value of recovery and reconstruction needs to be $11.5 billion.
- The Obama Administration requested $2.8 billion in FY2010 supplemental funding for relief and reconstruction support for Haiti.
- Extra-constitutional rule will begin after May 10, when most parliamentarians' terms expire; President Préval will probably rule by decree after that. Haiti passed a State of Emergency Law on April 20 giving the President broad powers. The international community is urging Haiti to hold legislative and presidential elections in November 2010.
- Humanitarian assistance is focused on providing waterproof emergency shelter, improving sanitation and meeting the basic needs of the displaced and other vulnerable Haitians. The relief effort is expected to last for many months.
- Most U.S. military forces will withdraw from Haiti by about June 1, 2010.

CURRENT CONDITIONS

The largest earthquake ever recorded in Haiti devastated parts of the country, including the capital, on January 12, 2010. The quake, centered about 15 miles southwest of Port-au-Prince, had a magnitude of 7.0. A series of strong aftershocks have followed.[1] The damage was severe and catastrophic. Thousands of buildings collapsed, leaving unknown numbers of people trapped, and hundreds of thousands of people homeless in the streets. Estimates of casualties are constantly being updated. According to the Secretary-General of the United Nations, "[o]f Haiti's 9 million people, initial reports suggest roughly a third may be affected by the disaster."[2]

The Inter-American Development Bank (IDB) issued a study on February 11 estimating the damage caused by the earthquake to be at least $8 billion, and as high as $14 billion.[3] One of the report's authors, economist Andrew Powell, said that, "This disaster, given the size of Haiti ... is the most devastating catastrophe that a country has experienced possibly ever."[4] The report compared the damage to other catastrophes, finding, for example, that the earthquake's damage amounted to 117% of Haiti's annual economic output, while the 2004 tsunami damage amounted to 2% of Indonesia's annual economic output. The Haiti Post Disaster Needs Assessment, conducted by Haiti and international institutions, reported similar findings, estimating the total value of recovery and reconstruction needs to be $11.5 billion.[5]

All of this damage was sustained in a country that the United Nations had already designated as one of the 50 "least developed countries" in the world, facing a higher risk than other countries of failing to come out of poverty, and therefore needing the highest degree of attention from the international community.[6]

Recovery efforts have been made extremely difficult by the loss of personnel and infrastructure that would be part of a recovery effort. Among the missing and dead were Haitian government officials and international aid personnel, including many U.N. personnel. Housing, hospitals, schools, and many government buildings collapsed. Basic services such as electricity and water were almost completely disrupted. Major transportation routes were damaged and/or blocked. The Port-au-Prince airport control tower was destroyed; the airport continued to function, and air traffic control authority was quickly transferred to U.S. personnel with portable radar. On February 18, commercial air traffic in and out of Port-au-Prince resumed. As of early March, relief flights into Haiti had dropped from a peak of 160 flights per day to an average of 75. The main port suffered heavy damage; U.S. troops are repairing the port facilities. The port is expected to handle 1,500 containers of relief supplies once it is fully repaired. The use of airfields and ports in the Dominican Republic, with humanitarian corridors over land, have also eased the burden on Haitian facilities.

Haitian government officials continue to function in makeshift conditions. The roof of the Presidential Palace collapsed and the President's private residence was also destroyed. President Préval was initially unable to communicate with his Cabinet and has been working out of a small room in a police headquarters. The Parliament building collapsed, with some Members trapped inside and others killed. The Parliament has convened in the National Police Academy. Twenty-eight of 29 Ministry buildings were destroyed.

The United Nations, which already had a strong presence in Haiti, is at the forefront of on-theground response for security and humanitarian assistance, and suffered heavy losses as well. Its headquarters collapsed, and 101 U.N. personnel are confirmed dead and 6 others remain unaccounted for. The head of the U.N. Stabilization Mission in Haiti (MINUSTAH), Special Representative Hedi Annabi, and his deputy, Luiz Carlos da Costa, were among the dead. U.N. Secretary General Ban Ki-moon sent Assistant Secretary General Edmond Mulet to Haiti on January 13 to direct the U.N.'s immediate response efforts; Mulet is a former Special Representative of the Secretary General for MINUSTAH.

Prior to the earthquake, the international community was providing extensive development and humanitarian assistance to Haiti. With that assistance, the Haitian government had made significant progress in recent years in many areas of its development strategy, including security; judicial reform; macroeconomic management; procurement processes and fiscal transparency; increased voter registration; and jobs creation. It had also made progress in providing broader access to clean water and other services. Parliamentary elections scheduled for February 2010 have been delayed. There is tentative agreement that Haiti will hold joint parliamentary and presidential elections in November 2010. President Préval's term will end in February 2011, and he is not eligible to run for reelection.

The destruction of Haiti's nascent infrastructure and other extensive damage caused by January's earthquake will set back Haiti's development significantly. U.N. Special Envoy and former President Bill Clinton said that Haiti's long-term development plans "will need to be amended ... but they cannot be abandoned."[7] Soon after the earthquake, the government of Haiti and the international donor community held meetings and conferences to begin planning a long-term strategy for Haitian recovery.

A major global donors' conference was held March 31, 2010, in New York at which 48 countries, multilateral institutions, and a coalition of non-governmental organizations pledged $5.3 billion for the first 18 months of Haiti's reconstruction, part of nearly $10 billion pledged toward longterm reconstruction efforts. The IDB damage assessment stated that donor coordination will be key in any reconstruction effort. The report further noted that there are different models for coordination, but that however it is coordinated, "[a] single executing agency with appropriate powers, transparency and accountability to the Haitian Government and donors would be helpful" in achieving the needed coordination on the ground.[8] To that end, Haiti and the international donor community have established a Multi-Donor Trust Fund to monitor, coordinate, and distribute funds.

The Obama Administration is requesting $2.8 billion in FY2010 supplemental funding to cover costs associated with relief and reconstruction support for Haiti. The U.S. pledge of $1.2 billion made at the donors conference is included in that request.

Preliminary Numbers at a Glance

It is estimated that 3 million people, approximately one-third of the overall population, have been affected by the earthquake. The government of Haiti is reporting an estimated 230,000 deaths (2% of the population), 300,600 injured and 383 missing.[9] Earlier reports indicated there were 1.2 million Internally Displaced Persons (IDPs), which included 700,000 people displaced in the Port-au-Prince area, many without shelter, and more than 597,000 people who left Port-au-Prince for rural areas.[10] More recent numbers indicate a higher number of IDPs with a possible total of over 2 million displaced in Haiti. This number includes 1.69 million displaced in settlements in Port-au-Prince and more than 300,000 in areas outside the city. Much smaller numbers of Haitians have left for the Dominican Republic, nearby islands, and the United States.

Aftershocks Pose Future Risk

A series of aftershocks has followed the main January 12 earthquake. In a February 23, 2010 statement, the USGS indicated that the frequency of aftershocks should diminish with time, but the threat of additional damaging earthquakes remains. The USGS noted that buildings in the Port-au-Prince area will continue to be at risk from strong earthquake shaking, and that the fault responsible for the January 12 earthquake still stores sufficient strain to be released as a large, damaging earthquake during the lifetime of structures built during the reconstruction effort. The USGS gave the following probabilities for aftershock activity in Haiti:

- magnitude 5 or greater earthquake = 90% probability over 1 year, 80% over 90 days, 55% over 30 days;
- magnitude 6 or greater earthquake = 25% probability over 1 year, 15% over 90 days, 7% over 30 days;
- magnitude 7 or greater earthquake = 3% probability over 1 year, 2% over 90 days, 1% over 30 days.

The USGS based its probability estimates on techniques developed to assess earthquake hazards in the United States. Using these techniques, the USGS estimated that the probability of a magnitude 7 or greater earthquake occurring within the next 50 years along the Enriquillo fault near Port-au-Prince is between 5% and 15%. The range of probabilities reflects the current understanding of the seismicity and tectonics of the Haiti region. By comparison, the USGS has estimated that that the probability of a magnitude 7 or greater earthquake occurring within the next 50 years along the Hayward-Rodgers Creek fault east of San Francisco is about 15%.

The USGS further asserted that the risk to new construction in Haiti from future aftershocks and strong earthquakes can be minimized through the use of earthquake-resistant design and construction practices. Tailoring the design to match the earthquake hazard likely requires applying results from a comprehensive earthquake hazard assessment, which currently does not exist for Haiti. This is further complicated by the need for buildings to withstand high winds from hurricanes, which are a more frequent hazard to Haiti than earthquakes. Constructing buildings that are both hurricane and earthquake resistant could pose a challenge to short- and long-term rebuilding efforts in Haiti.

Building construction is very important to earthquake resilience. For example, the magnitude 8.8 earthquake that struck Chile on February 27, 2010, caused only a fraction of the fatalities caused by the Haiti earthquake, in part due to the greater prevalence of earthquake-resistant buildings in Chile.

Despite the occurrence of two devastating earthquakes in Haiti and Chile within less than two months of each other, worldwide earthquake activity in 2010 is not particularly unusual compared to other years. The USGS reported that an average of 16 magnitude 7 or greater—defined as major—earthquakes have occurred each year since 1900, with some years showing considerable variation. Six major earthquakes have occurred worldwide in 2010 thus far, well within the normal range according to the USGS.

Sources: USGS statement, "USGS Updates Assessment of Earthquake Hazard and Safety in Haiti and the Caribbean," February 23, 2010, at http://www.usgs.gov/newsroom/article.asp?ID=2413&from=rss_home; and USGS release. "Is Recent Earthquake Activity Unusual? Scientists Say No," April 14, 2010, at http://www.usgs.gov/newsroom/article.asp? ID=2439.

According to the State Department, a total of 16,000 Americans in Haiti have been accounted for, and 16,704 U.S. citizens have been evacuated. There are 103 reported deaths of U.S. citizens due to the earthquake. Now that commercial airlines are operating out of Port-au-Prince, the U.S. embassy is no longer assisting with citizen evacuations.

HAITIAN GOVERNMENT RESPONSE

In the immediate wake of the earthquake, President Préval described conditions in his country as "unimaginable," and appealed for international assistance. The country's top priority was to conduct search and rescue operations for survivors. Other material priorities included an offshore vessel medical unit and electricity generation capability. The government also requested communications equipment so that government officials could better function and coordinate response efforts. As those immediate needs are met and the humanitarian relief operation continues, the government is struggling to restore the institutions needed for it to function and to address long-term reconstruction and development planning. "The first thing is political stability," said Préval. "Secondly, we hope the international community will help us in the short-term, midterm, and long-term."[11]

Prior to this disaster, the World Bank and others were working with the Haitian government to incorporate disaster risk management into Haiti's overall development strategy and to develop its capacity for disaster response. The capacity was still in its early stages, however, and the focus of much of its risk management efforts was not geared toward earthquakes, but toward hurricanes, which are the most common cause of natural disasters on the island. The last major earthquake in Haiti was 150 years ago, in 1860.

Some Haitians complain that they have seen or heard little of President Préval since the earthquake. The perception that the Haitian government was not doing enough contributed to calls by some Haitians, both in Haiti and the United States, for the United States to take control of Haiti in place of the current government in the days following the disaster.

The destruction of buildings, equipment, and loss of skilled personnel has drastically reduced the ability of the government to respond rapidly. According to the Haitian Chamber of Commerce, the earthquake destroyed approximately 25,000 public and commercial buildings. Twenty-one out of 22 Ministry buildings were destroyed. Along with the buildings, government records were destroyed; re-establishing and expanding transparency in government spending will be particularly challenging. These losses, plus the difficulty of delivering and transporting material supplies, will hinder delivery of services. The already significant need for services is now vastly expanded.

Haitian ministries are addressing issues such as long-term housing for those left homeless by the earthquake as they operate out of makeshift offices. Haitian authorities and international relief agencies are delivering food and water to hundreds of makeshift camps in Port-au-Prince. The government provided free transportation to evacuate people from the capital to cities not damaged by the earthquake. Interior Minister Paul Antoine Bien-Aime indicated that as many as 482,000 people may be relocated outside Port-au-Prince.[12] The Haitian government is sending officials to small cities to help officials in those communities establish priorities.

Other elements of the government are working along with international actors. The Haitian National Police are contributing to maintaining security, for example, and Haitian air traffic controllers are working along with U.S. controllers at the Port-au-Prince airport.

The Haitian government, the United Nations, and donor representatives met in Haiti on January 14 to coordinate their efforts, and have continued to do so at various conferences. Eighteen Haitian senators elected two commissions on January 28 to monitor aid and manage agreements with aid organizations.

The Government's New Action Plan

President Préval and Prime Minister Bellerive went to the March 31 international donors'conference with a 10-year recovery action plan. According to the Haiti Post Disaster Needs Assessment conducted by Haiti and international institutions, the total value of recovery and reconstruction needs is $11.5 billion.[13] At the conference The Préval Administration sought a minimum of $3.9 billion for the first 18 months of Haiti's reconstruction.

The priorities the Haitian government established in the "Action Plan for the Reconstruction and National Development of Haiti"[14] are:

- ensuring preparedness for the 2010 rainy and hurricane seasons, especially for displaced populations;
- providing assistance to the population affected by the earthquake while hastening recovery efforts with an eye to reducing dependence on foreign aid;
- including environmental factors and risk and disaster management in all recovery and reconstruction activities;
- actively providing employment and vocational training;
- strengthening governmental authority while also decentralizing basic services, creating a social safety net for the poorest population; and
- creating growth poles to support and encourage settlement of displaced populations around the country and to make Port-au-Prince less congested.

Haitian government officials see the earthquake's devastation in Port-au-Prince as an opportunity to establish a more sustainable distribution of population elsewhere in the country. Bellerive has asked international donors to aid in a massive revival of agriculture, in order to provide jobs and keep relocated Haitians in the provinces.[15]

Some observers have questioned whether historical and current allegations of various levels of corruption in the Haitian government may impair short- and long-term recovery efforts.[16] The government of Haiti made major progress in recent years in reducing corruption, increasing transparency, and improving fiscal management. These improvements qualified Haiti for Heavily Indebted Poor Country (HIPC) debt relief last year. In its new Action Plan, the government said the momentum of those programs must continue, and proposed the Interim Commission for the Reconstruction of Haiti, a Development Agency, and the Multi-Donor Trust Fund to improve development planning, coordination, and accountability. New concerns about transparency and oversight may also arise as the legislature ceases to function as a whole because elections have not been held. Normally, the legislature must approve federal procurement contracts, and authorize spending. Under a State of Emergency law passed by the legislature in April, the executive branch will have those powers for the next 18 months.

Extra-Constitutional Rule Begins after May 10

Haiti passed a State of Emergency Law on April 20 giving the President broad powers. The terms of all of the Chamber of Deputies, and of one-third of the Senate, will expire on May 10. How the government will function after that is not entirely clear. President Préval will probably rule by decree. The remaining 19 Senators may continue to meet, however, and the Senate has the power to dismiss the Prime Minister and censure government programs. As a result, political tensions are likely to increase. The international community is urging Haiti to hold legislative and presidential elections in November 2010. Préval has promised to step down when his term expires on February 7, 2011.

U.N. Stabilization Mission in Haiti (MINUSTAH)[17]

The U.N. Security Council created the U.N. Stabilization Mission in Haiti (MINUSTAH) on April 30, 2004, having determined that the situation in Haiti continued to be a threat to international peace and security in the region and acting under Chapter VII of the U.N. Charter. As a U.N.- conducted peacekeeping operation, MINUSTAH was given a mandate under three broad areas: a secure and stable environment, the political process, and human rights. On October 13, 2009, the Council extended its mandate until October 15, 2010, "with the intention of further renewal." The Council monitors the activities of MINUSTAH through semiannual reports made by the U.N. Secretary-General and his special representative, and also not later than 45 days before expiration of its mandate.

On January 19, 2010, the U.N. Security Council increased the overall force levels of MINUSTAH "to support the immediate recovery, reconstruction and stability efforts." The Council decided that "MINUSTAH will consist of a military component of up to 8,940 troops of all ranks and of a police component of up to 3,711 police and that it will keep the new levels of troops and police in MINUSTAH under review as necessary."[18] The limits had been set at 6,940 for the military component and 2,211 for the police component. As of April 13, 2010, the military component of MINUSTAH totaled 8,306 troops, including four engineering companies, and 2,178 police officers.[19]

On April 28, 2010, the U.N. Security Council met to consider the situation in Haiti, especially the April 2010 report of Secretary-General Ban Ki-moon on MINUSTAH.[20] This was his first report since September 1, 2009 and covered up to April 15, 2010, with a focus on events since the earthquake. Ban noted, in his report,

> that his Special Representative will continue to oversee the activities of the whole United Nations system and, within means and capabilities, ensure the full support of the MINUSTAH military, police and logistics components to humanitarian and recovery efforts.

Ban proposed that MINUSTAH "assume the lead role in coordinating international electoral assistance in order to ensure its efficiency and avoid the duplication of efforts." Ban observed that a "major focus of effort will be timely preparations for the 2010 presidential, legislative and municipal elections." [21] In addition, he stated that MINUSTAH "will reconfigure the military component within the existing troop ceiling to provide additional military engineers on a surge basis."[22] Ban also recommended that the Council increase the MINUSTAH police component, adding 680 police personnel to the force levels already authorized by the Council in January 2010.[23] As of May 3, 2010, the Council has not acted in response to the report of the Secretary-General.

The headquarters of the U.N. Stabilization Mission in Haiti (MINUSTAH) was demolished in the earthquake, with 101 U.N. staff confirmed dead. These included the Special Representative of the Secretary-General and Head of MINUSTAH Hedi Annabi and his Deputy, Luiz Carlos da Costa. On March 31, 2010, Secretary-General Ban announced the appointment of Edmond Mulet, as his Special Representative (SRSG) for Haiti and Head of MINUSTAH. Since January 14, 2010, Mr. Mulet had been acting SRSG and Head of MINUSTAH.[24] On April 29, 2010, Ban announced the appointment of Kevin Kennedy (United States) as Deputy Special Representative for Haiti and Deputy Head of MINUSTAH, succeeding Mr. da Costa, and Nigel Fisher (Canada) as Deputy Special Representative, Ad Interim, for Haiti, succeeding Kim Bolduc (Canada).

On March 24, 2010, President Obama, in a supplemental budget request to Congress on Haiti, requested in the State Department's Contributions to International Peacekeeping Activities (CIPA) account a total of $96.5 million for U.S. assessed contributions to MINUSTAH. This was to fund the increase in assessments resulting from the Security Council's increase in force levels on January 19, 2010. The March 24 supplemental request also included, under Foreign Operations, in the International Narcotics Control and Law Enforcement Affairs (INCLE) account, $45,000,000 to support U.S. personnel to MINUSTAH: the "secondment of 50 police advisers, five corrections advisers, and one drug specialist." These funds would increase the U.S. police and corrections officer contribution to MINUSTAH from 50 and 5 to 80 and 10, respectively.[25]

Humanitarian Relief Operation

Overall Status of the Relief Effort

Experts break relief operations into several phases: search and rescue; treatment and survival; relocation and rehabilitation; early recovery; and long-term reconstruction. As with any significant natural disaster that has many moving parts, it can take days to get a relief effort underway. Delays in transportation and congestion, lack of transportation infrastructure, bureaucratic problems, and lack of access all can cause bottlenecks at key points in the system. While timing is critical to save lives, to enable a network of this size to function efficiently requires the coordination of assessments and appropriate responses with the government, local communities, and the international community.

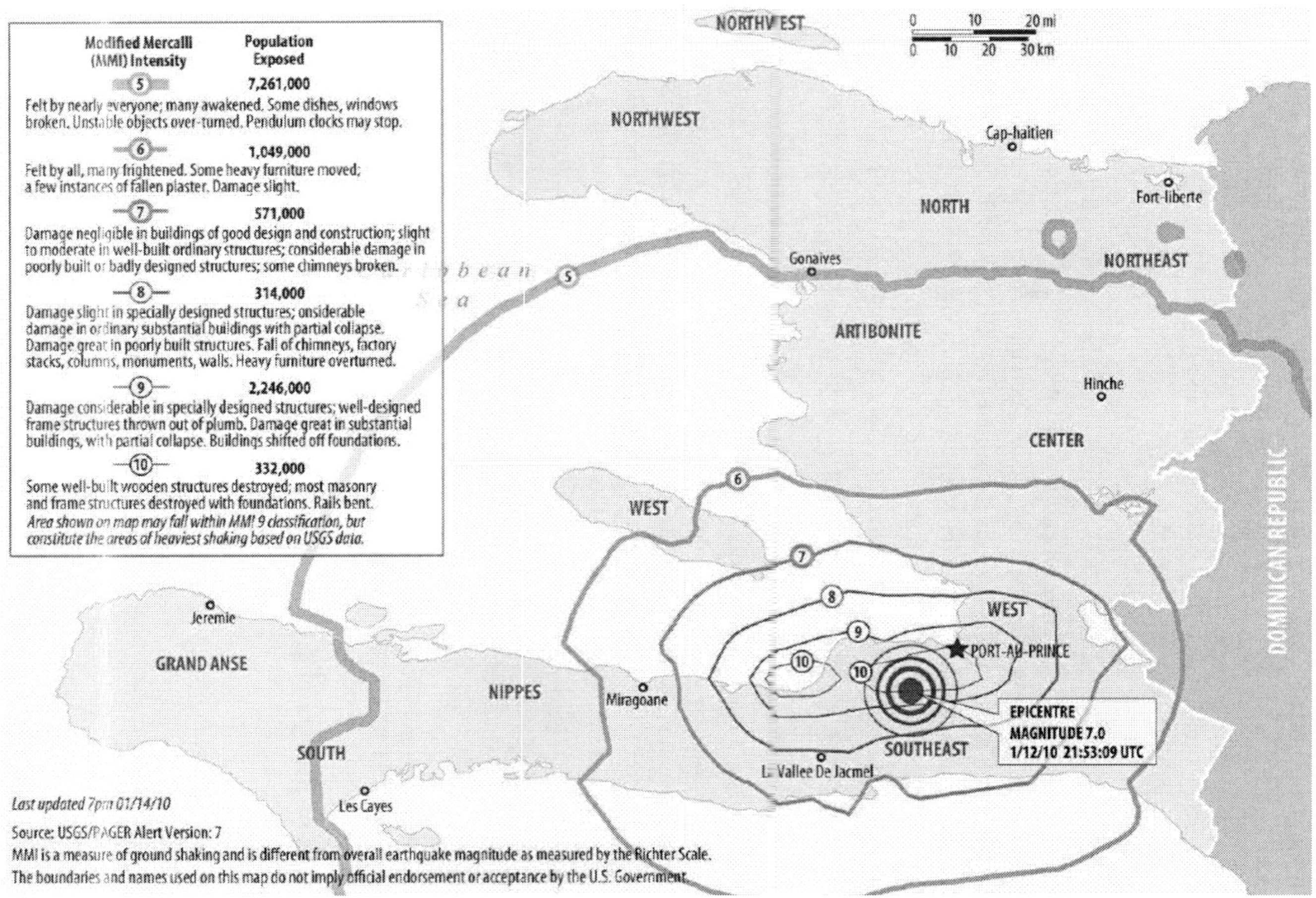

Figure 1. Haiti Earthquake Epicenter.

The sheer scale of the relief effort in Haiti has brought together tremendous capacity and willingness to help, but an ongoing effort and strategic planning is required at each phase to work out coordination and logistics issues. The massive humanitarian relief operation underway was hampered by a number of significant challenges, including a general lack of transportation, extremely limited communications systems, and damaged infrastructure. In many parts of Port-au-Prince, roads were ruptured or blocked by collapsed buildings, debris, bodies, and people seeking open space. Working conditions remain hugely challenging for aid personnel. Humanitarian supplies are coming in to Haiti via Port-au-Prince and Santo Domingo, Dominican Republic. The airport in the Dominican Republic is also being used as a humanitarian staging area to help with the coordination effort and allow for relief teams and supplies to get to Haiti by land.

Challenges consistent with a response to a disaster of this scope continue. In the first two weeks following the earthquake, priorities were focused on (1) search and rescue assistance, including teams with heavy-lift equipment and medical assistance and supplies; (2) addressing a critical need for food, clean water and sanitation, medical assistance, and emergency shelter; and (3) setting up key infrastructure and logistics operations. Three months after the disaster, the relief effort remains an immediate and critical priority amid plans under discussion for early recovery and reconstruction. There are concerns about security and potential for looting and violence, but according to the United Nations, the overall situation remains calm and stable, with some incidences of looting and criminality. However, 5,000 or so prison inmates escaped the National Penitentiary and other jails in Port-au-Prince during the earthquake (with some re-apprehended) and the whereabouts of the others is unknown, although it is suspected that a significant number may have retreated to Cite Soleil. Reports of rape cases in camps, roadblocks and fake tolls along roads, and other possible gang activity related to drugs and other crimes point to involvement by some of these criminals. U.N. and government officials are urging Haitians to turn in escaped criminals while also focusing on specific security interventions focused on their recapture.

More in-depth assessments, necessary to obtain a better understanding of the situation on the ground, continue. The Post-Disaster Needs Assessment (PDNA) a joint effort among representatives from the Government of Haiti, World Bank, United Nations and other actors, was completed in time for the March 31 conference in New York. The information is critical for determining whether personnel are in place with adequate resources, planning recovery and reconstruction initiatives, developing strategies for the use of funding, and paving the way for donor pledges and long-term support. The PDNA and donors conference are discussed later in the report.

United Nations Humanitarian Response

At the outset of the disaster, the United Nations established Disaster Assessment and Coordination (UNDAC) and U.N. Office for the Coordination of Humanitarian Affairs (OCHA) teams.[26] The UNDAC team coordinated the Onsite Operations and Coordination Center (OSOCC). Two sub-OSOCCs were established in Jacmel and Leogane to assist local authorities. The UNDAC team concluded its work in Haiti in February.

In the weeks following the earthquake, OCHA helped to coordinate the search and rescue teams and continues to coordinate the assistance effort while focusing on other humanitarian priorities. In addition to working closely with the government of Haiti, OCHA is the lead agency working with actors on the ground, coordinating with the military, and enlisting donor support. The Humanitarian Country Team convened on February 1 and meets twice a week, with at least one of those meetings co-chaired by a representative from the government of Haiti. In consultation with MINUSTAH and international military forces, OCHA has developed a Joint Operations Tasking Centre (JOTC) which began operating on January 26 and focuses on civil-military coordination and logistics. The OCHA Civil-Military Coordination (CMCoord) team first convened on January 31 and brings together civil-military points of contact from humanitarian organizations, MINUSTAH, and international military forces. A Debris Management Planning Task Force was assembled on February 23.

Humanitarian relief sectors are typically established during humanitarian crises to enable the United Nations to coordinate partners, prioritize resources, and facilitate planning. In Haiti, relief sectors have been organized into twelve clusters led by various agencies.[27] The clusters include

- Agriculture (Food and Agriculture Organization, FAO);
- Camp Coordination and Camp Management (International Organization for Migration, IOM);
- Early Recovery (U.N. Development Program, UNDP);
- Education (U.N. Children's Fund, UNICEF);
- Emergency Shelter and Non-Food Items (International Federation of Red Cross and Red Crescent Societies, IFRC);
- Emergency Telecommunications (World Food Program, WFP);
- Food (World Food Program, WFP);
- Health (World Health Organization, WHO, and Pan American Health Organization, PAHO).
- Logistics (WFP);
- Nutrition (UNICEF);
- Protection (Office of the High Commissioner for Human Rights, OHCHR), with Child Protection (UNICEF);
- Gender Based Violence (U.N. Population Fund, UNFPA);
- Water, Sanitation, and Hygiene (UNICEF).

Initially, there were criticisms of the slow initial response by the humanitarian aid community to the crisis. In recent press conferences, the United Nations has stated the cluster system appears to be working well, but others suggest that some clusters appear to be doing better than others.

Cluster arrangements have also been established in Leogane, Jacmel, and Gonaives. The cluster set up in the Dominican Republic was dissolved on April 13.

Humanitarian Relief Sectors: Recent Developments[28]

- **Emergency Shelter:** After the earthquake struck, people began gathering spontaneously in open spaces in Port-au-Prince. It was estimated that there were 500 or more makeshift camp sites with a combined total of more than 700,000 people.

Aid workers are delivering basic necessities to areas with population concentrations. Emergency shelter was in very short supply. Reports indicate that 97,000 houses were destroyed and 188,000 sustained damage. It is estimated that 1.9 million lost their homes.

Approximately 597,000 people were initially reported to have relocated in departments outside the city, with the highest number concentrated in Artibonite Department. Reports indicated that a number of areas saw increases of 15%-20% in the population. Ninety percent of the new arrivals were staying with host families. Reportedly, prices of basic commodities increased. See the maps in **Appendix B** and **Appendix C.**

In early April, the estimated number of displaced was adjusted, with an increase in the number of IDPs in Port-au-Prince to 1.69 million and IDPs displaced in areas outside the city appearing to decrease to 300,000. The adjustment may in part reflect more accurate numbers as a result of IDP registrations. It may also reflect a "pull factor" where those who left Port-au-Prince initially are now returning to the capitol as services improve and recovery gets underway.

The United Nations reports that 86% of the known caseload of IDPs have been reached with emergency shelter materials. As the rainy season begins, there remain concerns about whether there is sufficient waterproof emergency shelter, the need to move people from unsafe settlements, and ongoing challenges to improve sanitation. In addition, houses and buildings weakened by the earthquake could collapse or slide in wet conditions, adding to the 65 million tons of rubble needing to be removed. Finding space to shelter the displaced remains an issue.

For those living in unsafe areas, the International Organization for Migration (IOM) has identified several options: (1) return to the permanent home if the structure is considered safe; (2) return to home area to a temporary shelter or proximity site; (3) seek accommodation with host communities—usually relatives or friends; (4) relocate to a new site. The first phase of humanitarian relocations for vulnerable families living in high-risk displacement sites was completed at the end of April, with more than 1,750 families moving to two temporary relocation sites at the edge of Port-au-Prince. Another plot in Corail Cesselesse (Port-au-Prince) has received over 3,000 people.

IOM also reported that an estimated 49% of IDPs in the Port-au-Prince area were renters prior to the earthquake and 37% owned their own homes (14% did not identify their status.) Providing sustainable, long-term shelter to renters may be more problematic as renters do not own land for shelter installation and landowners may decide not to rebuild rental homes. In addition, owners of private schools have forced the eviction of those inhabiting their property as they seek to reopen their schools.

Where to build on a more permanent basis will be a major decision—many of the poorest people were squatters on land subject to landslides and floods. The type of structures to be built will also have to be determined as there will be a need for inexpensive housing that is both earthquake- and hurricane-resistant.

- **Food:** Food and water are reaching people every day, but the needs remain constant. It is estimated that 2 million people in Haiti need regular food aid. Some areas outside Port-au-Prince that were previously considered food secure are facing difficulties as prices increase. Supplementary food programs have been launched for children under 5 and pregnant women in the Port-au-Prince area. WFP has provided emergency food rations to 3.5 million people since the earthquake struck.

The heads of three U.N. agencies – WFP, the Food and Agriculture Organization (FAO) and the International Fund for Agricultural Development (IFAD) formed a new Task Force for Food Security in Haiti. The task force aims to assist the Haitian government with the implementation of a coordinated and targeted immediate and long term food security strategy. It will also work with the relevant U.N. humanitarian clusters. The WFP is now transitioning to a focus on long-term food security and investments in human capital programs. As part of this initiative, it is also rolling out food- and cash-for-work initiatives.

- **Health:** Acute respiratory illness is the most commonly reported illness, and in recent weeks, there have been some possible cases of malaria identified, but disease remains at baseline rates with none at epidemic levels. The health cluster is coordinating closely with other clusters on food assistance, shelter, sanitation, and education to assist the Haitian population. The need for sanitation and medical assistance remains. Limited follow-up, post-operative care is available and amputees are among those requiring longer-term assistance.

  The government of Haiti has shifted away from emergency services to focus on primary health care, health centers, and hospitals. The focus of the health cluster is also on basic primary health care services, such as maternal child health, rehabilitation services, and chronic diseases including diabetes, heart disease, HIV, and tuberculosis. Vaccination programs are being implemented and a targeted immunization program for populations in temporary settlements began on February 2. The provision of mental health care and psychosocial support to help survivors deal with trauma is also a recognized priority and organizations are mobilizing to address this issue.

  The identification and collection of mortal remains continues.
- **Logistics:** In both Haiti and the Dominican Republic, WFP is heading the in-country Food aid, Logistics and Emergency Telecommunications clusters. UNHAS is running a twice daily flight from Santa Domingo to Port-au-Prince. The WFP is prepositioning food and non-food items as well as equipment throughout Haiti in preparation for the hurricane season.
- **Protection:** The sub-cluster is focused on child protection activities in settlements, orphanages, and hospitals to determine the needs of children and to provide care to separated and unaccompanied children. Another main priority is protection of children during the relocation process. With so many people displaced and in need of protection, general security against crime is a critical concern. In addition, protection activities are focused on Haitian children displaced in the Dominican Republic.
- **Water, Sanitation and Hygiene (WASH):** A water supply strategy is being planned to phase out water trucking and to develop more sustainable options. Limited water network capacity remains a challenge. Access to safe drinking water has improved. Latrine usage and sanitation remain problems at spontaneous settlements. Increasing sanitation support is a main priority. This is seen as an important public health issue to avoid spread of disease, particularly during the rainy season. Plans are underway to increase the provision of water and sanitation services in communities.

- **Education:** Schools in areas unaffected by the earthquake reopened on February 1. Schools continue to open progressively, although the official reopening of schools was on April 5. Approximately 700 schools have now reopened in Port-au-Prince. The United Nations reports that 4,000 students were killed in the earthquake. It is estimated that 1,300 schools were destroyed and between 2,000 to 4,600 schools were damaged by the earthquake. Temporary schools are being provided with shelters, safe water, sanitation facilities, and educational materials for the rainy/hurricane season. Fourteen percent of the population of Haiti is under 14 years of age.
- **Early Recovery:** More than 95,700 people have been employed under UNDP's cash-for-work program. Activities include clearance of drainage canals and the removal of rubble.

Haiti's Humanitarian Needs in Context

A number of natural disasters have struck Haiti in the last decade (mostly in the form of hurricanes), prompting the need for international assistance through the United Nations and other actors. In addition to miscellaneous humanitarian funding provided for general emergency conditions, specific natural disasters include:

- Floods—May 2002
- Tropical Storm Jeanne/Floods—September 2004
- Floods—November 2006
- Hurricane Dean—August 2007
- Hurricane Felix—September 2007
- Tropical Storm Noel—October 2007
- Hurricanes Gustav and Ike, and Tropical Storms Fay and Hanna—September 2008
- Haiti Earthquake—January 2010

Disaster risks in Haiti are significant. Experts recognize that finding ways to overcome the cycle of disaster and develop a disaster response capacity are critical not only to minimize humanitarian consequences but to sustain reconstruction efforts in the future.

Other Humanitarian Actors

The international community has provided significant humanitarian assistance in response to these disasters and their ongoing impact.[29] The United Nations, along with other partners, including the United States, has had a strong presence in Haiti, and remains at the forefront of the current on-the-ground response for humanitarian assistance. Many international actors are providing humanitarian relief to Haiti, either through financial contributions to the government of Haiti or aid organizations or by directly providing relief supplies and emergency personnel.

International recovery efforts are typically complex because they require coordination among numerous different actors and international entities. In the current crisis, apart from U.N. agencies, those responding to humanitarian crises include international organizations, NGOs, Private Voluntary Agencies (PVOs), and bilateral and multilateral donors.

The International Federation of the Red Cross and Red Crescent Societies (IFRC) is working with the Haitian Red Cross Society (HRCS) and other national red cross societies, including the American Red Cross, to provide assistance to earthquake survivors. The IFRC is coordinating efforts with the International Committee of the Red Cross (ICRC), which is focused on medical assistance, tracing the missing and helping to restore family links. The ICRC is also helping with the identification and collection of mortal remains. In recent weeks, ICRC delegates have also visited 700 or so detainees in detention facilities in Port-au-Prince and Cap Haitien. Representatives of more than 20 national societies gathered in Montreal, Canada in mid February for a two-day meeting to develop a comprehensive approach with the red cross movement to Haiti's needs.

Various international NGOs that were already operating in Haiti before the earthquake mobilized to respond to the crisis. According to the United Nations, there are reportedly more than 900 NGOs focused on humanitarian relief operating in Haiti. Hundreds of local staff are assisting with the relief effort.[30]

U.S. Humanitarian Assistance

U.S. Humanitarian Assistance to Haiti FY2009-FY2011, Prior to the Earthquake Prior to the January earthquake, and in response to other natural disasters, the United States was already providing humanitarian assistance to Haiti (see Table 1 below.)

Table 1. U.S. Humanitarian Assistance to Haiti FY2009-FY2011 Appropriation by year, $ in thousands

	FY2009 Actual	FY2010 Estimate	FY2011 Request
TOTAL HUMANITARIAN ASSISTANCE	69,798	45,500	35,500
Public Law 480 (Food Aid)	63,767	45,500	35,500
International Disaster Assistance (IDA)	6,031	—	—

Source: U.S. Dept. of State, Office of the Director of U.S. Foreign Assistance, March 4, 2010.
Notes: Figures do not include Department of Defense assistance.

U.S. Humanitarian Assistance to Haiti following the Earthquake

Following the earthquake, on January 13, 2010, U.S. Ambassador to Haiti Kenneth H. Merten issued a disaster declaration, and the U.S. Agency for International Development (USAID), through the Office of Foreign Disaster Assistance (OFDA), authorized $50,000 for the initial implementation of an emergency response program. (See **Appendix E** for further information about the U.S. Government humanitarian response mechanism.) The embassy also facilitated the evacuation of U.S. citizens and issued a travel warning.

The U.S. government immediately set up an interagency task force to coordinate and facilitate the humanitarian response to the earthquake in Haiti through the Washington, DC-based Response Management Team (RMT) headed by U.S. Agency for International Development (USAID), through the Office of Foreign Disaster Assistance (OFDA). In the first three months following the earthquake, the overall focus of the U.S. government's response has included search and rescue, logistics and infrastructure support, provision of assistance, and conducting needs assessments.

On January 14, 2010, President Obama announced $100 million in humanitarian assistance (in addition to pre-existing funding appropriated for Haiti) to meet the immediate needs on the ground. As of April 30, 2010, USAID reported that the United States has provided more than $1 billion in humanitarian funding for Haiti as follows:

Table 2. FY2010 Humanitarian Funding Provided to Haiti for the Earthquake ($US)

Agency/Account	Amount
USAID/OFDA	382,395,438
USAID/FFP	110,263,500
USAID/OTI	35,000,000
USAID/Haiti	63,391,212
USAID/Dominican Republic	3,000,000
DOD	460,000,000
Total	1,059,050,150

Source: USAID/OFDA, Haiti-Earthquake, Fact Sheet #52, FY2010, April 30, 2010.

FY2010 Supplemental Humanitarian Relief Funding for Haiti

The Administration is requesting a total of $1.5 billion in relief and disaster assistance funding for Haiti, which would reimburse U.S. government agencies for services provided and for funds already obligated for ongoing relief activities. The humanitarian relief funding request also covers other relief-related assistance. The $1.5 billion request includes $350.7 million for USAID International Disaster Assistance (IDA); $150 million for Agriculture Department emergency food assistance; $96.5 million for State Department: Contributions to International Peacekeeping Activities; $655 million for Department of Defense and $45 million for U.S. Coast Guard relief activities; $220 million for Department of Health and Human Services to provide grants to States to cover services to Haitian evacuees; and $15 million for Department of Homeland Security immigration fees.

The activities of two of the key agencies—USAID and DOD—are described briefly below.[31]

USAID

Within 24 hours of the earthquake, the United States began deploying search and rescue teams along with support staff, and including search and rescue canines and rescue equipment, from Fairfax, VA, Los Angeles, and Miami. USAID/OFDA also deployed a 32-member Disaster Assistance Response Team (DART). The RMT (mentioned above) supported the USAID/DART, which focused on assessing humanitarian needs, positioning emergency relief supplies, and coordinating assistance with the U.S. Embassy in Haiti, the government of Haiti, and the international community. USAID personnel have been active in the following U.N. clusters: Water, Sanitation, and Hygiene; Emergency Food Assistance and Food Security Planning; Logistics; Health; and Shelter, and Protection. USAID/OFDA issues regular situation reports assessing the progress of relief operations.[32] See map in Appendix D. On April 25, the USAID/DART transitioned to the USAID/OFDA Program Office.

Department of Defense: Operation Unified Response[33]

In response to the crisis in Haiti, the Department of Defense (DOD) deployed a broad range of military assets in Operation Unified Response to support U.S. and international assistance efforts. On February 1, 2010, Admiral Mullen, chairman of the Joint Chiefs of Staff, stated that "We will remain in Haiti just as long as we are needed. At the request of the Haitian government and in partnership with the U.N. and international community, we will continue to do all that is required to alleviate suffering there." (See Appendix F for further information on the military units participating in Operation Unified Response.) At the peak of operations, DOD had deployed over 22,000 personnel, 15 Navy and Coast Guard vessels, and 58 aircraft to Haiti, including 7,000 ground troops. Currently, 1,000 soldiers remain ashore in Haiti and will remain through May, when they will be replaced with 500 Louisiana National Guard troops dedicated to reconstruction efforts. The USS Iwo Jima, an amphibious assault ship, will remain in the region through the upcoming hurricane season.

U.S. Southern Command (SOUTHCOM), located in Miami, oversaw the Department's response efforts, and U.S. Northern Command detailed over 100 disaster recovery specialists to Operation Unified Response. SOUTHCOM is well-experienced in this type of operation, having supported 11 relief missions in the Latin American and Caribbean area since 2005. SOUTHCOM's initial assessment team, consisting of military engineers, operational planners, and command and control communication specialists, deployed to Haiti within 24 hours of the earthquake. U.S. Air Force Special Operations Command personnel dispatched to the Port au Prince International Airport restored air traffic control capability and rapidly enabled round-the-clock airfield operations. The airport can now handle up to 120 flights a day, up from the 20 daily flights it handled prior to the earthquake. According to SOUTHCOM, over 14,000 U.S. citizens were evacuated safely.

U.S. military forces have delivered over 2.6 million bottled waters, 2.3 million food rations, 15 million lbs of bulk food, more than 125,000 lbs of medical supplies, and more than 844,000 lbs of bulk fuel. Additional tasks undertaken by DOD personnel included casualty treatment both ashore and afloat, aerial reconnaissance to assist rescue/supply efforts, the distribution of 73,000 hand-crank commercial radios, and the provision of radio broadcast capacity for emergency services information.

The U.S. Air Force's Air Mobility Command (AMC) provided a range of transport aircraft, including C-17 Globemaster IIIs and C-130 Hercules. Air National Guard units from Ohio and Puerto Rico also provided transport aircraft. The Navy Expeditionary Combat Command deployed units that can provide explosive ordnance disposal, maritime and riverine security, diving/salvage experts, and naval construction personnel. U.S. Navy surveillance and Air Force unmanned aircraft performed aerial surveys of the earthquake damage to assist remediation efforts. Of particular importance to improving rescue/recovery supply operations, a U.S. Coast Guard Maritime Transportation Recovery Unit coordinated efforts to restore the Port au Prince harbor facilities, and the U.S. Navy deployed a variety of specialized ships (salvage, heavy-crane, and oceanographic survey) to assist in these efforts. To date, eight Haitian ports are fully operational, and Port-au-Prince facilities are operating at partial capacity.

The 22nd Marine Expeditionary Unit (2,000 personnel) and a brigade combat team from the 82nd Airborne Division (3,400 personnel) have conducted security/humanitarian operations. Though there have been incidents of violence and looting, military commanders noted these have been concentrated primarily in areas known for violence prior to the earthquake, and the commanders are optimistic that violence will not spread to the general population, provided that the distribution of basic humanitarian supplies continues to improve. Fewer than 1% of the U.S. military personnel in Haiti have been required to conduct security operations. A U.S. Marine Fleet Anti-terrorism Security Team has provided protection for U.S. Embassy facilities. Civil Affairs units from the Army, Navy, and Marine Corps deployed to stricken areas to facilitate interaction between the population, the Haitian government, and organizations undertaking relief and recovery operations. The U.S. Coast Guard has undertaken the air-medical evacuation of injured U.S. civilian personnel to the Guantanamo Naval Station, supplied two C-130 transport aircraft, and deployed six cutters.

The Department of Defense has requested a supplemental appropriation of $655 million for its humanitarian relief efforts in Haiti. This request comprises $133.3 million each for both the Army and Navy Operations and Maintenance (O&M) Accounts, $133.4 million for the Air Force Operations and Maintenance, and $255 million for the Overseas Humanitarian, Disaster, and Civic Aid Account (OHDACA). The requested funds for the Services will back-fill the $400 million in O&M funds that were initially reprogrammed to the OHDACA account, the account used by DOD to conduct humanitarian relief efforts. The $255 million requested for OHDACA fulfills the remainder of the $655 million requirement for Operation Unified Response. This supplemental request covers expenses already incurred, in addition to permitting continued operations through June 15, 2010. Specifically, the funding will provide for soldier subsistence; personal, operational, transportation support; humanitarian relief supplies; and a number of humanitarian relief projects.

Non-Governmental Organizations (NGOs)

There is no central database that tracks activities and funding of national or international NGOs. It is therefore not possible to provide comprehensive and accurate information on their programs and funding, either for this crisis or for activities over the past decade. The United States Institute of Peace estimates that prior to the earthquake, the number of NGOs operating in Haiti ranged from 3,000 to 10,000.[34]

U.S.-based NGOs are playing an active role in the relief and recovery effort in Haiti, several of them with U.S. government funding. A list of U.S. NGOs working in Haiti can be obtained from a variety of sources.[35] A NGO Coordination Unit has been established to ensure better coordination among NGOs, the United Nations, and the military.

Overall U.S. FY2010 Assistance

In the FY2010 Consolidated Appropriations Act (P.L. 111-117) Congress provided "not less than $295,530,000" for assistance for Haiti, about $2.7 million more than the Administration had requested. Congress also included Haiti in the Caribbean Basin Security Initiative, to provide equipment and training to combat drug trafficking and related violence and organized crime, and for judicial reform, institution building, education, anti-corruption, rule of law activities, and maritime security. (See "Legislation in the 111th Congress" section below.)

The estimated FY2010 assistance for Haiti is $363 million, including $23 million and $121 million for Global Health and Child Survival under USAID and State Department, respectively; $161 million in Economic Support Funds; $35.5 million in P.L. 480 food aid; $21 million for International Narcotics Control and Law Enforcement; $0.22 million for International Military Education and Training, and $1.6 million in Foreign Military Financing.

The Administration had requested $359 million in FY2011 assistance for Haiti, including $35 million and $121 million for Global Health and Child Survival under USAID and State Department, respectively; $146 million in Economic Support Funds; $35.5 million in P.L. 480 food aid; $19 million for International Narcotics Control and Law Enforcement; $0.22 million for International Military Education and Training, and $1.6 million in Foreign Military Financing.

FY2010 Supplemental Funding Request for Haiti

The Obama Administration is requesting $2.8 billion in FY2010 supplemental funding to cover costs associated with relief and reconstruction support for Haiti following the earthquake.

The Administration has requested that all of the proposed funds be considered as emergency requirements, in response to urgent and essential needs in Haiti. Some of the funds are available until September 30, 2012, others until expended. The supplemental request covers both reimbursement of obligations already incurred and new activities by various U.S. agencies. CRS estimates that 55% of the total Haiti supplemental request is for reimbursement of relief activities related to the earthquake disaster, 40% for new recovery and reconstruction activities, and 6% for diplomatic operations administration.[36]

INTERNATIONAL HUMANITARIAN FUNDING

A great many international actors are also providing relief to Haiti, either through financial contributions to the government of Haiti or aid organizations or by directly providing relief supplies and emergency personnel.

U.N. Consolidated Appeals Process

The earthquake disaster in Haiti has received worldwide attention and focus. Under the U.N. Consolidated Appeals Process, on January 15, 2010, the U.N. Humanitarian Country Team in Haiti issued a Flash Appeal for emergency financial assistance in the amount of $575 million to support emergency food aid, health, water, sanitation, emergency education, and other key needs and early recovery efforts (typically the initial six months after a disaster). As of February 15, commitments of $619 million had been received (107 % of the Flash Appeal) and a further $29 million in uncommitted pledges. On February 19, 2010, the United Nations announced that it had revised its humanitarian aid appeal for Haiti to $1.44B – a record high. The Revised Humanitarian Appeal includes the original six-month Flash Appeal amount and it extends the humanitarian operation through 2010. As of April 29, 2010, commitments of $831 million (55.4% of the Revised Appeal) had been received and a further $27 million in uncommitted pledges.

Other Pledges and Contributions

Additional pledges and contributions have also been made outside the Flash Appeal. Many countries, including the U.S. government, are providing assistance in the form of direct contributions of items such as food and tents, or through the operation of relief flights and logistics support. In addition to bilateral assistance, funding has also been provided to NGOs operating outside of the U.N. appeal.

The Emergency Relief Response Fund for Haiti has more than $76 million in pledges, of which $63 million has been received. These funds have been allocated by cluster to U.N. and NGO organizations.

The U.N.'s Central Emergency Response Fund (CERF) initially made available $10 million which later increased to $37.9 million.[37]

Donor Contributions and Pledges

So far, through governments and the private sector, the international community has pledged millions of dollars in aid, materials, and technical support. **Appendix G** highlights donor contributions and in-kind pledges. At least 116 countries from around the world have contributed to the relief effort. Obtaining an exact up-to-date record of all international contributions is not possible—in part because some assistance is not reported to governments or coordinating agencies—and in part because of the delay in their recording. As the recovery effort gets underway, the Haitian government is considering a geographical distribution of tasks or tasks that are theme based as a way of engaging donor interest. For example, one country might "adopt" a province or city or target a specific need in the Haitian government.

Private Contributions

Private sector assistance has already been substantial and is expected to continue to grow. Some reports indicate that as of April 2010, private companies and individuals had contributed more than $1 billion to support relief efforts in Haiti.[38] Initiatives in the United States include the campaign by the American Red Cross to raise funds through text messages ($31 million), the Hollywood star-studded telethon that featured performances by a broad range of musicians and was broadcast on major U.S. television networks ($66 million), and numerous local fund raising activities.

On January 16, 2010, President Obama announced that former Presidents George W. Bush and Bill Clinton, who is also serving as the U.N. Special Envoy to Haiti, would lead a fundraising effort and work with the U.S. private sector in support of Haiti. The initiative is called the Clinton Bush Haiti Fund.[39] Cash donations are being encouraged. To date, the fund has received more than $200,000 in contributions. On March 22, Presidents Bush and Clinton visited Haiti to assess relief and recovery efforts.

Recovery Planning among Donors and Haiti[40]

President Préval has asked the international community to focus not just on immediate humanitarian relief efforts, but also on long-term development needs.

Early Recovery Planning

Discussions among the government of Haiti and the international donor community regarding a long-term strategy for Haiti began almost immediately. To that end, at a preliminary meeting among some international donors held in the Dominican Republic the week following the earthquake, Dominican President Leonel Fernandez proposed a $10 billion five-year assistance program for Haiti.

Representatives from Haiti, the "Friends of Haiti" nations, other countries, and U.N. officials held a high-level Ministerial Conference in Montreal, Canada, on January 25, 2010, to discuss reconstruction plans for Haiti. Haitian Prime Minister Jean-Max Bellerive thanked the donor community for its help so far, but said that an international commitment of 5 to 10 years was needed to support Haitian development. Conferees agreed to study recent examples of multilateral recovery efforts in order to develop an optimal aid-delivery mechanism that ensures effectiveness and accountability, and creates the conditions for sustainable development.

The World Economic Forum launched a global initiative to integrate business into Haiti's reconstruction at its meeting January 27-31.

March 31, 2010 Donors Conference[41]

The global donor conference in New York on March 31 secured commitments for substantial funds for Haiti's recovery. Donors pledged nearly $10 billion toward long-term reconstruction efforts, $5.3 billion of that for the first 18 months. These pledges are in addition to commitments made to the U.N. flash appeal for Haiti.

The Obama Administration, other international donors, the Haitian government, and others actors have all stated the need for improved accountability of all donor assistance to Haiti, to improve aid effectiveness and reduce the potential for corruption. Several entities have been established to improve development planning, coordination, and accountability. Although some Haitians think foreign donors are exerting too much control through these mechanisms, Haitian President Rene Préval and Prime Minister Jean-Max Bellerive have acknowledged that Haitian capacities were already limited, and were considerably diminished by the earthquake, and proposed such entities in their action plan. The National Assembly approved the Interim Haiti Recovery Commission.

Multi Donor Trust Fund

The World Bank will receive international funding and act as the fiscal agent for the funds. The Haitian government will lead a Steering Committee that will decide which agencies will receive the funds. The Bank will ensure that the Haitian government is meeting legal requirements before funds are transferred to it for budget support.

Haiti Reconstruction Platform

To ensure further transparency, the U.S. Agency for International Development has helped Haiti establish an internet-based system to track donor pledges, and monitor the implementation and spending of assistance.[42]

Interim Haiti Recovery Commission

Before the conference, the Préval Administration announced the formation of this Commission to guide long term development. It is to be co-chaired by former President Clinton and Haitian Prime Minister Jean-Max Bellerive. Other members are to include two Haitian legislators, local authorities, union and business representatives, a Caribbean Community (Caricom) delegate, and one representative of each of Haiti's largest donors. Those donors include the United States, Canada, Brazil, France, Venezuela and the European Union, as well as the Inter-American Development Bank, the World Bank, and the United Nations. It is designed to give Haitian officials the opportunity to exert leadership regarding their priorities, while working with international experts to hone their ability to design and execute development programs and process foreign assistance. According to Secretary of State Clinton's chief of staff, Cheryl Mills, the goal is to turn the process over to Haitian authorities in 18 months, at which time it would become the Haitian Development Authority.

The Role of the United Nations and Other Organizations

The United Nations, in association with other U.N. system agencies and programs, conducted the initial needs assessments necessary for planning Haiti's long-term and comprehensive recovery. Among the major actors involved were the World Bank, the U.N. Development Program, the multitude of U.N. specialized agencies (such as the World Health Organization, the U.N. Educational, Scientific, and Cultural Organization, the Food and Agriculture Organization), as well as regional organizations, including the European Union (EU), the Organization for American States (OAS), the Inter-American Development Bank, and the Caribbean Community (Caricom). Many of these organizations worked together previously in Haiti in response to the 2008 hurricanes and are responding to the international humanitarian needs in the aftermath of the earthquake. The Human Rights Council held a special session on Haiti on January 27-28, 2010, and in its resolution, it stressed the importance of protecting human rights during the recovery effort. The Office of the High Commissioner for Human Rights (OHCHR) has sent a human rights monitoring team to Haiti. The OHCHR along with the United Nations High Commissioner for Refugees (UNHCR) continue to press all countries to suspend involuntary returns to Haiti due to the humanitarian crisis.

RESPONSE OF INTERNATIONAL FINANCIAL INSTITUTIONS[43]

Multilateral Lending

The multilateral development banks (MDBs) have been active in Haiti for several years, providing debt relief, loans, and grants to both the Haitian government and the private sector. Following the earthquake, the World Bank, the Inter-American Development Bank (IDB), and the International Monetary Fund (IMF) announced new financial support for the country.

World Bank

After the earthquake, the World Bank announced $100 million in emergency grant funding to support recovery and reconstruction, in addition to its existing $308 million portfolio of grants projects in Haiti.[44] The existing projects are in areas including disaster risk management, infrastructure, community-driven development, education, and economic governance. At the March Donor's conference, the World Bank announced an additional $250 million in new funding for Haiti. This includes $151 million in grants, a $39 million write-off from cancelling Haiti's remaining debt to the Bank and $60 million in investments from the Bank's private sector arm, the International Finance Corporation (IFC).[45]

Inter-American Development Bank

On January 12, 2010, Inter-American Development Bank (IDB) President Luis Alberto Moreno announced a $200,000 emergency grant for immediate relief aid. The IDB is Haiti's largest multilateral donor, with a portfolio of programs worth over $700 million as of the end of 2009.[46] These programs include both grants and concessional loans. Of this amount, $330 million is undisbursed, of which $90 million could be quickly redirected to high-priority civil works and reconstruction projects.[47] IDB management also announced that it anticipates the approval of up to $128 million in already-planned grants, potentially providing more resources for reconstruction. In March 2010, IDB donor governments committed to providing necessary resources so that the IDB can provide at least $2 billion in grants to Haiti over the next ten years.[48]

International Monetary Fund

Haiti receives concessional loans from the International Monetary Fund (IMF) as well as from the multilateral development banks.[49] In response to the earthquake, the IMF approved $102 million in new assistance. Including the new lending, total Haiti debt to the IMF is now $271 million.[50]

Debt Relief Prior to the Earthquake

Haiti completed the multilateral Enhanced Heavily Indebted Poor Countries (HIPC) Initiative in June 2009, making it eligible to receive debt relief from the multilateral and some bilateral creditors.

World Bank

Under the terms of Haiti's participation in the Enhanced HIPC program, the World Bank provided Haiti debt relief for debts incurred through December 2003. According to the World

Bank, debt relief under the Enhanced HIPC Initiative amounts to $140.3 million. Inter-American Development Bank

The IDB, in September 2009, provided $511 million in debt relief. Debts eligible for cancellation were those incurred through 2004 (compared to 2003 in the case of IDA). According to the IDB, Haiti currently owes $429 million (principal-only) to the IDB. This includes $305 million from loans made in 2005 and 2006, after the debt cancellation cut-off date of December 31, 2004, and $124 million from undisbursed balances of loans made before the cut-off date. Beginning in 2009, Haiti's payments on its debt to the IDB have been made by a U.S.-supported trust fund that currently amounts to $20 million.

Bilateral Creditors

Haiti has also received debt relief from its bilateral creditors. Haiti's completion of the HIPC program triggered debt relief of $62.7 million by the Paris Club group of official creditors. Haiti's Paris Club creditors agreed to go beyond the requirements of the HIPC program, however, and provide $152 million in additional debt cancellation, thus completely cancelling Haiti's external Paris Club debt of $214 million. That said, Paris Club debt relief is not automatic. Creditor nations collectively sign bilateral agreements with the debtor nation, giving effect to the multilateral debt relief agreement. By mid-January, for example, France had only cancelled €4 million ($5.75 million) of €58 million ($83.36 million) owed to them by Haiti.[51]

On September 18, 2009, the United States cancelled $12.6 million (100%) of Haiti's outstanding debt to the United States. The United States has not extended new loans to Haiti since September 2009. Haiti does not currently have any outstanding debt obligations to the United States.

Debt Relief Since the Earthquake

Agreement has been reached by several donor nations to finance the cost of additional Haiti debt relief for three institutions: the IDB, the World Bank, and the International Fund for Agriculture and Development (IFAD).

Inter-American Development Bank

Haiti's remaining debt from the IDB totals $447 million. An estimated $186 million of additional loans have been approved but have not yet been disbursed. Once these are provided, they will add to the debt stock. There is also a $28 million balance that represents Haiti's local currency conversion obligation to the IDB's low-income lending arm, the Fund for Special Operations (FSO). At the IDB's March 2010 annual meetings, member states agreed to fund 100% cancellation of Haiti's remaining debts owed to the IDB, although all details have not been finalized.[52]

World Bank

Haiti's remaining debt to the International Development Association, the World Bank's low-income lending facility, is $39 million. On March 31, 2010, the World Bank announced that 14 donor countries, Belgium, Canada, Germany, Finland, France, Ireland, Italy, Japan,

The Netherlands, Norway, Spain, Sweden, Switzerland and the United States, have pledged to provide a total of $39 million to cancel Haiti's remaining IDA debt.[53]

International Fund for Agriculture and Development

Haiti's remaining debts to IFAD are around $79 million. In April 2010, member nations agreed to cancel these debts.[54] Under the agreement, IFAD - an international financial institution and a specialized United Nations agency dedicated to eradicating poverty and hunger in rural areas of developing countries - will contribute up to 30 per cent of the debt relief requirement, with member states needing to contribute the remaining 70%.

Since the earthquake, other countries have stepped forward to provide debt relief to Haiti. At the end of January 2010, Venezuela announced it would forgive Haiti's debt, totaling $295 million.[55] In mid-February 2010, French President Nicolas Sarkozy visited Haiti and announced a new French aid package for Haiti that reportedly included the cancelling of all the debt owed by Haiti to France, totally €56 million (approximately $77 million).[56] Taiwan, another major bilateral creditor to Haiti with $90 million of outstanding debt to Haiti, has discussed canceling Haiti's debt but nothing has materialized to date.[57]

International Monetary Fund

To date, there has been no international agreement on forgiveness of Haiti's debts to the IMF of $271 million. The interest rate on Haiti's IMF loans is zero until 2012, however, with no payments due until 2013.

The IMF and Haiti's Financial Sector

On February 24, 2010, the IMF issued a press release on its role in helping revive Haiti's financial sector.[58] Officials from the IMF and Haiti's Central Bank discussed a proposal prepared by Haiti's authorities to maintain financial stability and restart private sector credit. The IMF's press release states that "this proposal will be critical to allowing the private sector to fully play its role in rebuilding the economy and providing jobs, in the context of the broader reconstruction and economic recovery."[59] The IMF also announced that it will help Haiti's authorities design a partial credit guarantee fund.

Debt Relief in the FY2010 Supplemental Funding Request

To help Haiti in its recovery from the earthquake, the Administration is proposing U.S. contributions of $252 million to help cancel Haiti's debts of $781 million to three international organizations: the Inter-American Development Bank, World Bank, and International Fund for Agriculture and Development. Through FY2010 supplemental funding, the Administration requests reallocating up to $40 million from the Treasury Department's Debt Restructuring Account appropriated for the multilateral Heavily Indebted Poor Countries (HIPC) Trust Fund from this or subsequent fiscal years to Haiti. The Administration also seeks new contributions of $212 million for multilateral debt relief. Congressional authorization is required for the $40 million reallocation, and both authorization and appropriations are required for the additional $212 million.

REGIONAL RESPONSE

Latin American countries have responded to Haiti's crisis with immediate provision of emergency supplies and personnel and pledges of financial and other assistance for its long-term recovery. Members of the Organization of American States (OAS) pledged humanitarian, financial and other support to Haiti. The OAS Group of Friends of Haiti met on January 14 to coordinate search and rescue efforts, prompt donations, and discuss ways to promote recovery.

The 15-member Caribbean Community (CARICOM), of which Haiti is also a member, was working in Haiti prior to the earthquake. Afterwards, it mobilized its disaster emergency response system to assist Haiti, and several members have sent emergency supplies or promised financial assistance. The Caribbean Disaster Emergency Management Agency assembled a response team to assess conditions in Haiti as well.[60]

Many countries in the region already have peacekeeping troops in Haiti serving with MINUSTAH. Brazil leads the U.N. peacekeeping mission, and had 1,284 uniformed personnel already serving there as of December 2009.

Many countries in the region have made bilateral cash or in-kind contributions as well. (See "Regional Donors to Haiti Relief Effort," in Appendix F.)The Dominican Republic, which shares the island of Hispaniola with Haiti but did not suffer heavy damage from the earthquake, responded swiftly and generously. The two countries have a long history of hostility toward one another, but Presidents Préval and Fernandez have worked in recent years toward having a more cooperative relationship, and this has been reflected in the Dominican response. Haiti's neighbor was the first country to send relief supplies and personnel, and has facilitated aid delivery through use of its airports, roads, and port. It has stopped repatriation of undocumented Haitians, and opened its border to injured Haitians, thousands of whom have been treated in both public and private hospitals. Fernandez organized a preparatory meeting for donors to discuss future aid to Haiti the week after the earthquake, and the Dominican Republic hosted a technical meeting prior to the March 31 donor conference.

POLITICAL AND ECONOMIC SITUATION IN HAITI[61]

Conditions in Haiti Before the Earthquake

Long before the earthquake struck, Haiti was a country socially and ecologically at risk. It has some of the lowest socioeconomic indicators in the world[62]. Haiti was already in an acute environmental crisis. Only 2% of its forest cover remains intact.[63] Following the hurricanes of 2008, the President of the Inter-American Development Bank (IDB), Luis Moreno, called Haiti the most fragile of IDB's member countries, saying that no other nation in Latin America and the Caribbean is as vulnerable to economic shocks and natural disasters as is Haiti.

Haiti had been making progress, however. The U.N. Secretary-General commissioned a report, published in January 2009, that recommended a strategy to move Haiti beyond recovery to economic security. Indeed, the U.N. Security Council conducted a fact-finding visit to Haiti in March 2009, and concluded that there was "a window of opportunity to enable the consolidation of stability and the undertaking of a process of sustainable development." [64]

Political Conditions

President Préval is in his second (non-consecutive) five-year term as President of Haiti. During the first three years of this term, Préval established relative internal political stability. He outlined two main missions for his government: (1) to build institutions, and (2) to establish favorable conditions for private investment in order to create jobs. In November 2007, his Administration published its National Strategy for Growth and Poverty Reduction, a key step in meeting IMF requirements for debt relief, which it met in June 2009. With the support of MINUSTAH, which arrived in Haiti in 2004, security conditions improved, as did the capacity of the country's police force. Both the former and current U.S. Administrations praised Préval for his efforts to improve economic conditions and establish the rule of law in Haiti. Préval pledged to cooperate with U.S. counternarcotics efforts. Both President Barack Obama and Secretary of State Hillary Clinton met with President Préval early in 2009, and since the earthquake have provided humanitarian assistance and pledged long-term support for development in Haiti

The Haitian government is functioning under extremely difficult conditions, with 28 of 29 government Ministry buildings destroyed, and 17% of the country's civil service killed.[65] U.S. and U.N. officials both say they are coordinating relief and recovery efforts with the Préval administration. To provide the Haitian government some operating space, the U.S. Department of State agreed on January 16 to lease the old U.S. Embassy building in downtown Port-au-Prince to the Haitian government for $1 a year. That building had been put up for sale in June 2008 after the new U.S. Embassy opened near the Port-au-Prince airport. On February 16, 2010, Canadian Prime Minister Stephen Harper while in Haiti announced that Canada would provide $12 million to build temporary facilities to house several Haitian government ministries.

Though greatly improved, Haiti's political stability remains fragile. Préval's inauguration in 1996 was the first transition between two democratically elected presidents in Haitian history. The government has its third prime minister since April 2008. Parliament dismissed Prime Minister Michele Pierre-Louis in October 2009, barely a year after her appointment. Nonetheless, the transition was smooth as President Préval swiftly appointed, and the Parliament confirmed, Jean-Max Bellerive to take her place. As Minister of Planning and External Cooperation from 2006 to 2009, Bellerive helped to prepare Haiti's National Strategy for Growth and Poverty Reduction.[66]

After the earthquake, former President Jean-Bertrand Aristide announced he would like to return to Haiti, although he has since given no explicit plans to do so. Aristide has lived in exile in South Africa since his government collapsed in 2004. Once—and possibly still—extremely popular among some Haitians, he is nonetheless a divisive figure. Aristide would face charges of corruption and would likely contribute to political instability if he were to return.

The National Assembly and the Question of 2010 Elections

Haiti's parliament, the National Assembly, also faces enormous challenges in trying to reestablish itself: some of its members were killed in the earthquake; the parliament buildings were destroyed, as was the electoral council's building. The U.S. House of Representatives' Democracy Partnership and others are working with the Parliament to help it function again. The National Assembly has been meeting since shortly after the earthquake in makeshift offices. The Senate elected two commissions on January 28 to monitor international aid and manage agreements with aid organizations.

Political tensions were mounting ahead of parliamentary elections scheduled for February 28 and March 3, 2010. In late 2009, President Préval cut ties to the Lespwa movement that elected him in 2006, and formed a new movement, Unity. Opposition groups accused the presidentially appointed electoral council of bias in favor of the President's new movement. The electoral council disqualified, without explanation about 15 rival political groups, which included members of Lespwa who did not join Préval's new party. Opposition groups expressed concern that if Unity won a legislative majority, it would push through constitutional amendments, possibly including one allowing Préval to run for another term in 2011. The constitution allows only two non-consecutive terms for a President, and Préval is completing his second term. Préval has promised to step down when his term expires on February 7, 2011, and said he would not run again. There is no clear candidate to succeed him. In February 2010, the electoral council postponed the elections indefinitely. The elections were to determine all 99 seats in the House and one-third of the seats in the Senate.

Tensions are likely to increase when those legislative terms expire on May 10, 2010 without replacements having been elected. As mentioned earlier, it is not entirely clear how the government will function after that date. But Haiti passed a State of Emergency Law on April 20 giving the President broad powers and President Préval will probably rule by decree. After May 10 the remaining 19 Senators may continue to meet, however, and the Senate has the power to dismiss the Prime Minister and censure government programs. The international community is urging Haiti to hold legislative and presidential elections in November 2010.

Socioeconomic Conditions Prior to the Earthquake

Plagued by chronic political instability and frequent natural disasters, Haiti remains the poorest country in the Western Hemisphere. Haiti's poverty is massive and deep. Over half the population (54%) of 9.8 million people live in extreme poverty, living on less than $1 a day; 78% live on $2 or less a day, according to the World Bank.[67] Poverty among the rural population is even more widespread: 69% of rural dwellers live on less than $1 a day, and 86% live on less than $2 a day. Hunger is also widespread: 81% of the national population and 87% of the rural population do not get the minimum daily ration of food defined by the World Health Organization. In remote parts of Haiti, children have died from malnutrition.[68]

In order to reach its Millennium Development Goal of eradicating extreme poverty and hunger by 2015, Haiti's Gross Domestic Product (GDP) would have to grow 3.5% per year, a goal the International Monetary Fund (IMF) says Haiti is not considered likely to achieve. Over the past 40 years, Haiti's per capita real GDP has declined by 30%. Therefore, economic growth, even if it is greater than population growth, is not expected to be enough to reduce poverty. Haiti has experienced some economic growth since 2004. Economic growth for FY2007 was 3.2%, the highest rate since the 1990s. Before the earthquake, the forecasted growth for FY2009-20010 was 2.5%, reflecting the impact of recent storms and the global economic crisis, and up to 3.5% for 2010-2011.[69] The global economic crisis also had led to a drop of about 10% in remittances from Haitians abroad, which in 2008 amounted to about $1.65 billion, more than a fourth of Haiti's annual income.[70]

The likelihood that economic growth will contribute to the reduction of poverty in Haiti is further reduced by its significant income distribution gap. Haiti has the second largest income disparity in the world. Over 68% of the total national income accrues to the wealthiest 20% of the population, while less than 1.5% of Haiti's national income is accumulated by the poorest 20% of the population. When the level of inequality is as high as Haiti's, according to the World Bank, the capacity of economic growth to reduce poverty "approaches zero."[71]

LONG-TERM RECONSTRUCTION STRATEGY

Review of Haiti's Development Strategy

Haiti already had a National Strategy for Growth and Poverty Reduction in place for 2007-2010, supported by the international donor community.[72] As Minister of Planning and External Cooperation from 2006 to 2009, current Prime Minister Jean-Max Bellerive helped to prepare that Strategy.[73] The poverty reduction strategy focuses on three "priority pillars." The first is areas for growth, focusing on agriculture and rural development, tourism, infrastructure modernization; and science, technology and innovation. The second pillar is human development, concentrating on education and training; health; water and sanitation; persons with disabilities; childhood poverty; young people; HIV/AIDS; and gender equity. The third pillar calls for investment in democratic governance, focusing on the establishment of an equitable justice system; creation of a climate of security; modernization of the state; and political and economic decentralization.

After a series of devastating hurricanes in 2008, the government of Haiti revised its strategy incorporating the findings of a Post-Disaster Needs Assessment and the U.N.'s "Haiti: From Natural Catastrophe to Economic Security" report, at a donors conference held April 21, 2009, in Washington, DC. The Haitian government outlined the priorities of its new two-year plan, "Haiti: a New Paradigm," which include investing in strategic infrastructure, improving economic governance and the business environment, improving the provision of basic services, and ensuring environmental sustainability.[74]

The government was making strides toward meeting goals of its growth and poverty reduction strategy, and some analysts were viewing its potential for sustainable development with optimism. Investors were returning to Haiti and the country was promoting its economic development. The earthquake has reversed years of progress. Haiti's strategy will therefore need to be reviewed, revised, and built upon to incorporate new conditions and needs.

Haiti's National Strategy for Growth and Poverty Reduction remains relevant, and the government says its pillars will remain the basic pillars of its strategy. The Préval Administration will present a revised vision statement at the March 31 donors conference, and more fully revise its strategy after the completion of the post-disaster needs assessment.

The U.S. Department of State was about to announce a new strategy toward Haiti, on which it had been collaborating with the Préval Administration for almost a year.[75] That, too, will need to be revised. The assessment concluded that a new strategy needed: (1) a comprehensive integrated approach to achieve sustainable long-term stability and economic growth; (2) investment in plans led by the Haitian government to ensure sustainability; (3) better coordination to maximize the effectiveness of U.S. and other donor assistance; (4) expanded reach of U.S. programs by using partnerships with other international actors; and (5) improved accountability and measurement of results.

The State Department's assessment and plan focused on four areas: agriculture, energy, health, and security.[76] All of these areas, plus others such as governance and education, will need to be addressed in the short term, while simultaneously developing plans to rebuild in the long term.

Principles of Assistance for the Reconstruction of Haiti[77]

At the March donors conference in April, the donor countries agreed to the following principles:

Ownership: Donors will align their assistance with priorities established by the Haitian government and will involve the Haitian people in the development and implementation of projects.

Coordination and Effectiveness: Participants will strengthen coordination through consultation and "unprecedented" transparency, and concentration on priority sectors for each donor, and will adhere to the principles of aid effectiveness and build on lessons learned.

Inclusiveness: The conference emphasized the importance of consulting with key communities, such as the Haitian people, civil society, non-governmental organizations, local governments, the private sector, the Haitian diaspora, MINUSTAH stakeholders, and regional partners.

Accountability and Transparency: The government of Haiti and donors agreed to increased accountability and transparency in both pledges and use of assistance, and established a monitoring and transparency unit within the Interim Haitian Reconstruction Commission/Haitian development authority accountable to the Haitian and international public.

Sustainability: The international community pledged a long-term commitment to Haiti's reconstruction.

Select Issues to be Addressed in the Long-Term Reconstruction Strategy

The priorities of the Haitian government's Action Plan for the Reconstruction and National Development of Haiti are outlined in "The Government's New Action Plan" earlier in the report. Analysts and donors are stressing that Haiti cannot be merely re-built, but must be re-built better. In this crisis, many people see the opportunity to address some of the underlying problems contributing to the country's endemic poverty and underdevelopment. The Action Plan addresses these issues, and is the basis for donor contributions and activity.

Agricultural Capacity and the Environment

Rebuilding Haiti's agricultural capacity is a key element of the country's reconstruction strategy and is seen as a way of broadening Haiti's economy, and reducing its reliance on food imports. Yet Haiti's environment was in a state of crisis before the earthquake struck. Obstacles to agricultural development include massive deforestation, erosion of topsoil, lack of investment in agricultural technology for decades, and unclear land titles. A relatively new infrastructure is in place to help with the process of incorporating post-earthquake needs into Haiti's overall environmental rehabilitation plan. Following the destructive hurricane season of 2008, efforts to revive Haiti's ecology had been renewed when a coalition of U.N. agencies, government agencies, non-governmental organizations, and technical institutes launched a Haiti Regeneration Initiative.[78]

Decentralization of Population and Services

A core goal of the revised Haitian development strategy, supported by the donors conference, is to catalyze economic growth and provide services and opportunities outside of Port-au-Prince. According to U.N. and U.S. officials, decentralization is the focus of the second phase of delivery assistance as well, and food, shelter, jobs programs, and other services are being delivered to communities around the country to help them cope with the influx of people from the capital.

Haiti was once a predominantly rural population, with only about 20% of its population living in cities. By the time of the 2010 earthquake, the vast majority of Haitians lived in cities, primarily Port-au-Prince. Parts of the Haitian government and private sector have concentrated resources, services, and job opportunities in Port-au-Prince for decades. Prime Minister Bellerive and analysts who follow Haiti suggest that the current crisis provides an opportunity to correct what had become an unsustainable urban-rural distribution of people and resources in the country. Some have suggested not rebuilding Port-au-Prince because it lies on a fault line and remains susceptible to further earthquakes.

Property rights are not clearly established throughout Haiti, and are of concern as the Haitian government and donors try to set up relocation settlements, reforestation and agricultural programs, and encourage investors to build outside of the capital. U.S. programs will help develop infrastructure in support of decentralization and help seek resolutions to property rights issues.

Education

Canada and France are the countries designated to focus on helping Haiti develop education in the reconstruction phase. Haiti's schools were woefully inadequate prior to the earthquake, and many were destroyed by the earthquake. Most schools are privately run. Education is crucial to raising Haitians out of a cycle of poverty, by providing the knowledge and skills individuals need to take advantage of job opportunities. Experts note that job creation must be accompanied by education programs.

Energy

Energy is one of the United States' areas of focus in the development strategy. By virtually all accounts, Haiti's current energy sources are inefficient and inadequate. They are often destructive as well: Haitians' reliance on charcoal for fuel has contributed to the deforestation of all but 2% of its forest cover. Some observers have suggested that clean energy technology could help Haiti "avoid some of the poverty traps of the old system."[79] According to at least one analysis, developing small-scale, alternative energy sources at the local level rather than trying to rebuild the previously ineffectual Haitian electricity service would increase the quality of life of many Haitians and have a positive impact on economic growth.[80]

Health Care

Because the United States is already a leading provider of health care in Haiti, it is one of the areas on which it will focus in the reconstruction phase. In much of the country the government did not provide basic services prior to the earthquake. The lack of medicines or medical treatment and adequate sanitation in Haiti has been exacerbated by the earthquake. In the long-term, health care is crucial to raising Haitians out of a cycle of poverty, by providing the good health that enables children to develop and adults to function fully, whether as students, family providers or employees.

Job Creation

Job creation in both the short- and long-term are important parts of the recovery strategy. UNDP has already launched cash-for-work programs both to stimulate the local economy and ease the delivery of humanitarian assistance. The program will quickly expand to earthquake-damaged areas in and outside of Port-au-Prince, and employ 220,000 people, indirectly benefiting about one million Haitians, according to UNDP.[81] The current jobs are for clearing streets of building rubble and disposing of debris. Donors are also creating job programs outside Port-au-Prince to benefit communities receiving an influx of people from the capital. Public works such as building roads, housing, and infrastructure are to generate jobs in the reconstruction phase.

Trade and Exports

Plans for economic growth include restoring and continuing to expand industrial exports. Many analysts emphasize, however, that economic plans must be comprehensive, to avoid over-reliance on any one area, such as the apparel assembly industry, which could leave the Haitian economy overly vulnerable.

Recovery and Reconstruction Funding in the FY2010 Supplemental Request

Within the Haiti supplemental proposal, the request for Recovery and Reconstruction funding is $1.1 billion. This is primarily for new activities, in support of the Haitian government's Action Plan. U.S. programs to be funded through the supplemental request focus on urgent infrastructure repairs, especially in the energy and agricultural sectors; critical health care; governance; and security.

Congressional Concerns

Many Members have already expressed a strong desire to support Haiti and provide it with substantial assistance. The 111th Congress gave bipartisan support to assist the Préval government during its first session, and has continued to respond in that spirit to the crisis generated by the January earthquake. Committees in the Senate and House have held hearings on Haiti. Both Members and witnesses stressed the need for a massive, coordinated international effort not only for immediate humanitarian needs, but also for long-term development. Moving forward, they said, strategies must consider new approaches, aim to create a more sustainable Haiti, and increase Haitian capacity to utilize foreign aid effectively and to provide services and direct its own economy.

FY2010 Supplemental Funding Request for Haiti

Regarding the Obama Administration's total request for $2.8 billion in FY2010 supplemental funding for costs related to Haitian relief and recovery activities, key concerns include:

Choosing Priorities

To coordinate aid programs better, donors have agreed to focus on certain areas of assistance. Some observers have expressed concern that U.S. assistance is neglecting areas crucial to Haitian recovery, such as improving the educational system, which is to be the focus of Canada and France. While advocates say this approach avoids duplication among donors, critics question the priorities, or the limited approach to aid.

Some experts suggest developing small-scale, alternative or clean energy sources at the local level rather than trying to rebuild the previously ineffectual Haitian electricity service would increase the quality of life of many Haitians and have a positive impact on economic growth.[82] Some Members have expressed concern that insufficient funding is being focused on the needs of children, or on psychological support for the traumatized population. There is no additional funding for Global Health and Child Survival in the supplemental request. Some long-term care for psychological support is included in the health activities proposed in the supplemental.

Decentralization and Economic Growth: Will they Lead to Poverty Reduction?

A key element of the revised Haitian development strategy, supported by the supplemental request, is to catalyze economic growth and provide services and opportunities outside of Port-au-Prince. The Haitian government and donors agree that the current crisis provides an opportunity to correct what had become an unsustainable urban-rural imbalance in the country, with the rest of the country suffering neglect while people, resources, and services were concentrated in the capital. Funds in the supplemental request would address both short- and long-term elements involved in this decentralization strategy – meeting the immediate needs of newly displaced populations that have migrated to less developed areas of the country, and strengthening local governance, infrastructure, and agriculture to develop new "growth poles" outside of Port-au-Prince. Scientists are helping Haitian authorities to select areas for development that are less vulnerable to natural disasters. While there is general support of this strategy, officials also note that developing areas long-neglected will be costly. Some also warn that populations should not be forcibly relocated in executing these plans. Experts also warn that economic growth is not sufficient to reduce poverty in Haiti, and that programs specifically targeted at poverty reduction are needed. As mentioned earlier, the

World Bank says that when the level of economic inequality is as high as Haiti's, the ability of economic growth to reduce poverty approaches zero.

Effective Capacity Building?

Most observers agree that one goal of aid to Haiti should be to build the capacity of Haitians so they can eventually assume responsibility for the project at hand. Yet there is a tension between the standard definition of effectiveness and efficiency, and the time and money required for capacity building. Aid organizations are pressed to have measureable outcomes and usually operate on short-term contracts. If thorough training and coordinating with Haitian ministries is to be an element of all foreign aid programs, which many experts advocate, there will have to be a recognition that those programs may require more time, funding, and personnel, and measureable results may take longer to achieve.

Budget Priorities

Humanitarian assistance generally receives strong bipartisan congressional support and the United States is typically a leader and major contributor to relief efforts in humanitarian disasters.[83] When disasters require immediate emergency relief, the Administration may fund pledges by depleting its disaster accounts intended for worldwide use throughout a fiscal year. President Obama announced the United States would provide $100 million in immediate aid for Haiti. That aid is drawn from existing funds. The international community is also making substantial donations toward meeting immediate needs.

Amid efforts to tackle rising budget deficits by, among other measures, slowing or reducing discretionary spending or finding the resources to sustain U.S. aid pledges may be difficult. After the 2004 tsunami disaster, some Members of Congress publicly expressed concern that funding for tsunami relief and reconstruction, which depleted most worldwide disaster contingency accounts, could jeopardize resources for subsequent international disasters or for other aid priorities from which tsunami emergency aid had been transferred. These accounts were fully restored through supplemental appropriations. At the time, others noted the substantial size of American private donations for tsunami victims and argued that because of other budget pressures, the United States government did not need to transfer additional aid beyond what was already pledged. In Haiti, the full extent and cost of the disaster is not yet known. Disaster accounts are being drawn down to provide relief to Haiti. The State Department reports that in order to respond to future humanitarian crises, these resources would need to be replenished by June 1, 2010. If not replenished, U.S. capacity to respond to other emergencies could be impacted. The relief funding in the current FY2010 supplemental request would provide reimbursement for funding already provided or obligated.

Congress may reevaluate and revise priorities and approaches of U.S. assistance to Haiti in light of the changed conditions there. Issues that have previously concerned Congress have included democracy building, development assistance and poverty reduction, security enhancement and stability, counternarcotics efforts, police and judicial reform, and disaster recovery and prevention.

Burdensharing and Donor Fatigue

The earthquake disaster in Haiti has received worldwide attention and focus. The government of Haiti, the United States, the United Nations and many others have asked for and encouraged governments to provide assistance. It is not always evident whether figures listing donor amounts represent pledges of support or more specific obligations.[84] Pledges made by governments do not necessarily result in actual contributions. It also cannot be assumed that the funds committed to relief actually represent new contributions, since the money may previously have been allocated elsewhere. It will take time for a more complete picture to reveal how the actual costs of the Haiti disaster will be shared among international donors. Comparing USG and international aid is also difficult because of the often dramatically different forms the assistance takes (in-kind contributions vs. cash, for instance). As the situation in Haiti stabilizes, and early recovery and reconstruction gets underway, sustaining donor interest in Haiti (and commitment to honor existing pledges) could be a challenge. Moreover, this challenge is compounded by the need to maintain funding priorities and secure funds needed for other disaster areas worldwide.

Coordinating the Relief Response in Haiti

Some have criticized the response by the international community in the actual delivery of humanitarian assistance as far too slow. For example, in the days following the earthquake some press reports commented on what they perceived to be a critical lack of food and water, insufficient medical care for the wounded, the slow pace of search and rescue, and the non-existent presence of law and order. Others have argued that there has been a great deal of unfair criticism of the slowness of the international aid effort. The weakened capacity of the Haitian government, critically damaged infrastructure, and logistical challenges posed by the influx of massive aid into a city largely destroyed by the earthquake all contributed to delay and difficulties on the ground.

Still others have been concerned about bureaucratic red tape, lack of civil-military cooperation, control by the U.S. military of flight priorities at the Port-au-Prince airport, and overall coordination issues. Evaluations of the relief response in Haiti will likely continue to be conducted and debated as the humanitarian effort moves ahead. A disaster of this scope is almost certain to run into many obstacles because the challenges on the ground are so daunting. While managing expectations of what is possible under these circumstances is important, so too, are the observations and lessons learned that with time and hindsight may benefit the actions and plans of those responding to future disasters. A U.N. Inter-Agency Real Time Evaluation (RTE) Mission was launched in April and will be conducted in several phases, the results of which will also be incorporated into the implementation of the Revised Humanitarian Appeal.

Elections in Haiti

Another issue of concern to Congress is likely to involve arrangements regarding elections. The Haitian Electoral Council is responsible for carrying out elections, and

MINUSTAH provides technical and logistical support. The United Nations is encouraging the Haitian government to get back on an election schedule as soon as possible, to avoid the possible politically destabilizing effect extended delays and rule by decree might create. The United States is encouraging Haiti to hold presidential elections when they are due to be held, in fall 2011, and to combine them with legislative elections.

There are currently enormous obstacles to conducting elections before year's end. During the earthquake, election experts were killed, and the U.N.'s offices were destroyed, along with records and equipment. Throughout the country, many buildings in which elections are usually held were destroyed. In addition, many Haitian citizens lost their identity cards, and many more have been displaced and will need to register at their new locales. Some elections experts say it is possible to carry the elections out this year if the preparations are begun immediately.

Immigration, Adoption[85]

The devastation caused by the January 12, 2010, earthquake in Haiti led Department of Homeland Security (DHS) Secretary Janet Napolitano to grant Temporary Protected Status (TPS) to Haitians in the United States at the time of the earthquake.[86] As soon as the earthquake hit, some Members of Congress had called for the Obama Administration to do so. On January 13, 2010, DHS had announced that it was temporarily halting the deportation of Haitians. On January 15, 2010, DHS Secretary Napolitano granted TPS to Haitian nationals for 18 months.

Haitian children who were legally confirmed as orphans eligible for intercountry adoption by the government of Haiti and who were in the process of being adopted by U.S. residents prior to the earthquake have been given humanitarian parole to come to the United States. Other Haitian orphans potentially eligible for humanitarian parole include children who were identified by an adoption service provider or facilitator as eligible for intercountry adoption and who were matched to prospective American adoptive parents prior to January 12, 2010. [87] The parole program stopped taking new applications on April 14, 2010. DHS anticipates that approximately 1,200 orphans will ultimately receive parole under this program.[88] The Haitian government says its first priority regarding displaced children is to try to reunite them with relatives.

On January 29, 2010, 10 U.S. citizens, most from an Idaho Baptist group, were arrested in Haiti and charged with child kidnapping and criminal association. The U.S. citizens were trying to cross the Haitian border with the Dominican Republic with a busload of 33 children who they said were orphaned by the January 12, 2010, earthquake. The U.S. citizens did not have authorization to remove the children from Haiti, and it appears that many of the children were not orphans.[89] On February 11, Haitian Judge Bernard Sainvil ordered the release of the U.S. citizens on their own recognizance "because there had been no evidence demonstrating 'criminal intentions' on their part to support charges of child kidnapping and criminal association leveled against them."[90] The charges have been dropped against all 9 of the U.S. citizens, while the remaining U.S. citizen will be tried in Haiti on the charge of arranging "irregular travel."[91]

Those Haitians who are given humanitarian parole to come to the United States are deemed Cuban-Haitian Entrants and, thus, are among the subset of foreign nationals who are eligible for federal benefits and cash assistance much like refugees. Those Haitians who are newly arriving legal permanent residents, however, are barred from the major federal benefits and cash assistance for the first five years after entry. The President has included funding to cover additional costs for federal benefits and cash assistance resulting from Haitians evacuees, among other activities in his FY2010 supplemental request. Specifically, the President has requested $220.0 million for the Department of Health and Human Services (HHS) to fund four types of activities, of which two are directly related to Haitians brought to the United States after the earthquake. [92] These two are: the state share of Medicaid and Children's Health Insurance Program (CHIP) costs for eligible Haitians; and cash, medical, and repatriation assistance for eligible Haitians. The supplemental request also includes $15 million for the U.S. Citizenship and Immigration Services to use for fee waivers for eligible Haitians granted TPS, those given humanitarian parole to bring medical evacuees and certain categories of Haitians into the United States, and to cover costs associated with processing the adoption of Haitian orphans.

According to the Department of State, there are 54,716 Haitians who have approved petitions to immigrate to the United States and who are waiting for visas to become available. Advocates for Haitians are asking Secretary Napolitano to give humanitarian parole to those Haitians with approved petitions for visas. Proponents of expediting the admission of Haitians with family in the United States maintain that it would relieve at least some of the humanitarian burden in Haiti and would increase the remittances sent back to Haiti to provide critical help as the nation tries to rebuild. Those opposed to expediting the admission of Haitians assert that it would not be in the national interest, nor would it be fair to others foreign nationals waiting to reunite with their families.

There are growing concerns that the crisis conditions in Haiti may result in mass migration from the country. The phenomenon of Haitians coming to the United States by boat without proper travel documents dates back at least to the 1970s. The Reagan Administration reached an agreement in 1981 with the Haitian government to interdict (i.e., stop and search certain vessels suspected of transporting undocumented Haitians), and this policy, with some modifications, has continued. During the first two weeks of February, the U.S. Coast Guard returned almost 90 Haitians found at sea to northern Haiti, although U.S. officials say there is no sign of a mass exodus from Haiti by sea.[93] If mass migration were to occur, Congress may weigh in on the balancing of immigration control responsibilities with humanitarian concerns in the midst of Haiti's humanitarian disaster.

Medical Evacuation[94]

In the initial weeks after the earthquake, U.S. military airlifts brought non-U.S.-citizen Haitians to Florida for medical care that was unavailable in Haiti. These patients, who often had very serious injuries, were admitted to a number of non-federal hospitals, principally in south Florida. On January 27, 2010, Florida Governor Charlie Crist wrote to Kathleen Sebelius, Secretary of the U.S. Department of Health and Human Services (HHS), saying that the state's health care system was reaching saturation, and asking Sebelius to activate the National Disaster Medical System (NDMS) to coordinate the distribution of medical evacuees to other states, and to compensate states for the costs of their care.[95] On February 1, 2010, HHS announced that it was activating the NDMS hospital component.[96] Federal medical evacuations resumed at that point, but are no longer underway as of early March, 2010,[97] as the most critical needs have abated or can now be addressed in Haiti.

HHS administers the NDMS, which consists of three components for the response to mass casualty incidents.[98] First, teams of medical providers deploy rapidly to provide critical medical care in disaster conditions when local health care capacity has been destroyed or overwhelmed.[99] Second, NDMS provides medical evacuation, the coordinated movement of seriously ill or injured victims to sites where they can receive definitive care. Typically, U.S. military assets such as cargo planes and attending medical personnel are used for this purpose. Third, participating U.S. hospitals agree to provide medical care to NDMS evacuees on a voluntary basis, and receive federal reimbursement at 110% of the Medicare rate.

The first NDMS component is used frequently. Teams are deployed many times each year in response to domestic disasters (for which deployment costs are typically paid by the Federal Emergency Management Agency's Disaster Relief Fund[100]) and international humanitarian relief efforts (for which deployment costs are often paid by USAID). NDMS teams were deployed to Haiti shortly after the earthquake. In contrast, the second (medical evacuation) and third (U.S. participating hospital) NDMS components have rarely been activated. Although U.S. military flights had initially brought Haitian medical evacuees to Florida during January, the NDMS hospital component had not yet been activated, and it was not clear at that time if the federal government would assume the costs of care for those patients.

The HHS Secretary has considerable discretion with respect to the activation of NDMS. No specific legal triggers or other requirements must be met.[101] However, the Secretary does not have a dedicated funding mechanism to support extensive NDMS deployments. Under HHS policy for this incident, NDMS will reimburse hospitals for the costs of care, for 30 days, for any individual who was medically evacuated from Haiti by NMDS, regardless of citizenship or nationality.[102] NDMS does not pay costs beyond 30 days, costs for services provided by non-hospital facilities (such as rehabilitation facilities), or costs for the care of individuals who were not evacuated through the NDMS system. President Obama has requested emergency supplemental funding to support the Haiti earthquake response. The request includes $220 million for HHS, which would be used, among other things, to reimburse NDMS hospitals that cared for medical evacuees.[103]

Tax Incentives for Charitable Donations

In a bipartisan effort, Congress passed a bill designed to increase charitable donations to Haiti. The Haiti Assistance Income Tax Incentive (HAITI) Act (signed into law January 22, 2010, P.L. 111-126) accelerates income tax benefits for charitable cash contributions for the relief of earthquake victims. It allows taxpayers to deduct donations made in early 2010 on their income tax returns for 2009. The Joint Committee on Taxation estimates that the HAITI Act would lead to U.S. revenue losses of about $2 million.[104] Information on how to make charitable donations is in Appendix I.

Trade Preferences[105]

One long-standing targeted policy that the U.S. Congress has taken to support economic development in Caribbean countries is to provide tariff preferences for their exports. Haiti is eligible for such benefits under the Caribbean Basin Trade Partnership Act (CBTPA—P.L. 106- 200) and the Haiti Hemispheric Opportunity through Partnership Encouragement (HOPE) Act, as amended (P.L. 110-246). In particular, flexible and generous tariff preferences are provided to Haiti's apparel assembly business, which has been established for decades, with admittedly mixed success over time, and viewed by many as one pillar of economic growth in an economy with limited opportunities in the short-run.[106]

The U.S. Congress continued to support this approach with the Haiti Economic Lift Program (HELP) Act (S. 3275), introduced on April 28, 2010. The bill, which had bicameral and bipartisan support, extends the CBTPA and the Haiti HOPE Act through FY2020. It focuses on expanding preferences for apparel exports, benefits some have argued might also be extended to other industries to help promote economic diversification. Currently, however, such opportunities are very limited, and Congress appears to have opted to review the HOPE Act carefully and enhance those trade preferences rules that have had the most demonstrable effect in stimulating increased apparel production and exports for the U.S. market, and by extension, job creation in Haiti.

Apparel accounts for 93% of Haitian exports to the United States and reflects a comparative advantage based on low-cost labor and proximity to the U.S. market. Earthquake damage to apparel firms created new hardships for this industry, and there is need for a considerable increase in both domestic and foreign investment for the apparel industry to survive and expand. Of the 23 plants operating in late 2009, the earthquake completely destroyed one, killing some 500 people, and seriously damaged four others. Currently 19 are fully operational, two are being relocated, and two are closed. Employment attendance rates have returned to levels seen prior to the earthquake, but with fewer factories operating, total employment has fallen from 26,500 to 23,300. Monthly apparel exports have declined from $58.2 million in February 2009 to $33.1 million in February 2010.

Focusing trade assistance on this sector attempts to respond to Haiti's level of development and play into the best prospects for creating employment in the near term. The strategy does have its critics, who argue that it has had limited effects in the past and should be employed within a broader development plan that seeks to diversify growth of output and exports over the longer term. In addition, critics contend that more needs to be done to provide better job opportunities as the country faces wholesale reconstruction. Congress recognizes these problems and has responded with broader development assistance and within the parameters of the HOPE Act with the detailed labor provisions. For example, the HOPE Act has put in place a system that requires Haitian firms that wish to avail themselves of the tariff preferences to conform to internationally recognized labor standards, and submit to an independent factory monitoring system operated by the United Nations.

Constituent Concerns

Lawmakers are also helping constituents find persons missing in Haiti. Information on how to help them do so is in Appendix H. Members were also helping citizens in Haiti leave

Haiti. Now that commercial air traffic has been resumed, there are no more emergency evacuations being carried out.

Lawmakers may also seek to find ways for the Haitian and U.S. governments to speed pending and potential adoptions of Haitian orphans. Links for further information on adoptions and orphans are in Appendix H.

Many constituents want to know how to contribute to relief efforts. Information on how to do so is in Appendix I.

LEGISLATION IN THE 111TH CONGRESS[107]

P.L. 111-8. In the Omnibus Appropriations Act, 2009, Section 7045 makes the government of Haiti eligible to purchase U.S. defense articles and services for its Coast Guard. It also obligates funds for (1) Haiti under Titles III and VI of this Act; health care, nutrition, sanitation, education, and shelter for migrant workers and others. It prohibits the use of specified funds under this Act for the transfer of U.S. weapons, ammunition, or other lethal property to the Haitian National Police until the Secretary certifies to the Appropriations Committees that any members of the Haitian National Police alleged to have committed serious crimes, including drug trafficking and human rights violations, have been suspended. Introduced February 23, 2009, signed into law March 11, 2009.

P.L. 111-117. In the Consolidated Appropriations Act, 2010, Sec. 7045(b) deals expressly with Haiti, stating that, (1) The government of Haiti shall be eligible to purchase defense articles and services under the Arms Export Control Act (22 U.S.C. 2751 et seq.), for the Coast Guard; (2) of the funds appropriated by this Act under Titles III, Bilateral Economic Assistance, and IV, International Security Assistance, not less than $295,530,000 shall be made available for assistance for Haiti; and (3) none of the funds made available by this Act under the heading "International Narcotics Control and Law Enforcement" may be used to transfer excess weapons, ammunition or other lethal property of an agency of the United States Government to the government of Haiti for use by the Haitian National Police until the Secretary of State reports to the Committees on Appropriations that any members of the Haitian National Police who have been credibly alleged to have committed serious crimes, including drug trafficking and violations of internationally recognized human rights, have been suspended.

Haiti is included in Sec. 7045(c), as part of the Caribbean Basin Security Initiative. The section reads as follows:

> Of the funds appropriated under the headings 'Development Assistance,' 'Economic Support Fund,' 'International Narcotics Control and Law Enforcement,' and 'Foreign Military Financing Program' in this Act, not less than $37,000,000 should be made available for assistance for the countries of the Caribbean Basin, to provide equipment and training to combat drug trafficking and related violence and organized crime, and for judicial reform, institution building, education, anti-corruption, rule of law activities, and maritime security, of which not less than $21,100,000 should be made available for social justice and education programs to include vocational training, workforce development and juvenile justice activities: *Provided*, That none of the funds made available under this subsection shall be made available for budget support or as cash payments.

The Act calls on the Secretary of State to provide a detailed spending plan to the Committees on Appropriations no later than 45 days after this Act is enacted, for funds appropriated or otherwise made available for the countries of the Caribbean Basin, with concrete goals, actions to be taken, budget proposals, and anticipated results. Introduced July 22, 2009, signed into law on December 16, 2009.

P.L. 111-126. The Act to Accelerate the Income Tax Benefits for Charitable Cash Contributions for the Relief of Victims of the Earthquake in Haiti, allows taxpayers who donate to Haiti earthquake relief between January 11, 2010, and March 1, 2010, to claim those contributions on their 2009 tax return. Introduced January 19, 2010, signed into law on January 22, 2010.

P.L. 111-127. The Emergency Aid to American Survivors of the Haiti Earthquake Act amends title XI of the Social Security Act (SSA) to increase the funding cap under the U.S. Repatriation Program to $25 million for FY2010 for temporary assistance to U.S. citizens (and their dependents) returning from foreign countries in the event of destitution, illness, war, threat of war, invasion, or similar crisis; amends SSA title XIX (Medicaid) to provide $65 million in additional funding for the Qualified Individual (QI) Program which pays the Medicare part B premium costs for low-income seniors. Reduces a corresponding amount otherwise available to the Medicaid Improvement Fund for FY2014. Introduced January 25, 2010, signed into law January 27, 2010.

P.L. 111-158. Debt Relief for Earthquake Recovery in Haiti Act of 2010. Amends the International Financial Institutions Act to direct the Secretary of the Treasury to instruct the U.S. Executive Directors at the International Monetary Fund (IMF), the International Bank for Reconstruction and Development, the Inter-American Development Bank, and other multilateral development institutions to: (1) cancel all debts owed by Haiti to such institutions; (2) suspend Haiti's debt payments to such institutions until the debts are canceled completely; and (3) provide additional assistance from such institutions to Haiti in grant form in order to avoid additional debt accumulation. Introduced February 2, 2010. Signed into law on April 26, 2010.

H.R. 144. The Haitian Protection Act of 2009 would require the Secretary of Homeland Security to designate Haiti as a country whose qualifying nationals may be eligible for temporary protected status. Introduced January 6, 2009; referred to the House Judiciary Committee's Subcommittee on Immigration, Citizenship, Refugees, Border Security, and International Law February 9, 2009.

H.R. 264. The Save America Comprehensive Immigration Act of 2009 would amend the Immigration and Nationality Act (INA) to provide increased protections and eligibility for family-sponsored immigrants, including to authorize adjustment of status for certain nationals or citizens of Haiti. Introduced January 7, 2009, referred to House Judiciary; House Homeland Security; House Oversight and Government Reform Committees; referred to the Subcommittee on Immigration, Citizenship, Refugees, Border Security, and International Law February 9, 2009.

H.R. 417. The Next Steps for Haiti Act of 2009 would authorize the Director of Foreign Assistance, in consultation with the government of Haiti and Haitian civil society organizations, to establish the Haiti Professional Exchange Program to assign qualified Haitian Americans and others to provide technical assistance to help Haiti improve in areas vital to its growth and development, including education, energy, environment, health care, infrastructure, security, transportation, and disaster preparedness. Directs the Secretary of

State to implement a student loan forgiveness program for program participants. Introduced and referred to the House Committee on Foreign Affairs January 9, 2009.

H.R. 1567. The Haitian Refugee Immigration Fairness Act (HRIFA) Improvement Act of 2009 would amend the 1998 HRIFA to (1) require determinations with respect to children to be made using the age and status of an individual on October 21, 1998 (enactment date of the HRIFA of 1998); (2) permit an application based upon child status to be filed by a parent or guardian if the child is present in the United States on such filing date; and (3) include document fraud among the grounds of inadmissibility which shall not preclude an otherwise qualifying Haitian alien from permanent resident status adjustment. It would also permit new status adjustment applications to be filed for a limited time period. Introduced March 17, 2009, referred to the House Judiciary Committee's Subcommittee on Immigration, Citizenship, Refugees, Border Security, and International Law on April 27, 2009.

H.R. 3077. The Global Food Security Act of 2009, partner legislation with **S. 384**, authorizes the President to provide assistance under this Act or the Foreign Assistance Act of 1961 for unexpected urgent food assistance needs. Establishes a United States Emergency Rapid Response to Food Crisis Fund to carry out such purposes. Introduced June 26, 2009.

H.R. 4206. The Haiti Reforestation Act of 2009 seeks to authorize the Secretary of Agriculture to provide assistance to the government of Haiti. The purpose of the act is to end the deforestation in Haiti within five years and to restore the tropical forest cover to its state in 1990 within a 30- year time frame. The legislation was both introduced and referred to the House Committee on Foreign Affairs on December 3, 2009.

H.R. 4468 Haiti Action Initiative and Tax Incentive Act of 2010. Treats cash contributions made during January 2010 for the relief of earthquake victims in Haiti as having been made on December 31, 2009, for purposes of the tax deduction for charitable contributions. Introduced on January 19, 2010 and referred to the House Committee on Ways and Means.

H.R. 4577. Emergency Health Services for Haitian Earthquake Victims Act of 2010. Introduced and referred to the House Committee on Energy and Commerce February 3, 2010.

H.R. 4603. Haitian Orphan Placement Effort Act or the HOPE Act – Would direct the Secretary of Homeland Security (DHS) to expand the humanitarian parole policy for certain Haitian orphans; would apply on a case-by-case basis to children who were legally confirmed as orphans eligible for inter-country adoption by the government of Haiti before January 12, 2010; would authorize the placement of Haitian children granted humanitarian parole into the United States in an unaccompanied refugee minor program if a suitable family member is not available to provide care. Introduced February 4, 2010 and referred to the House Committee on the Judiciary.

H.R. 4616. Haitian Emergency Life Protection Act of 2010. Would temporarily expand the V nonimmigrant visa category to include Haitians whose petition for a family-sponsored immigrant visa was approved on or before January 12, 2010. Introduced on Feb. 5, 2010 and referred to the House Committee on the Judiciary. Referred to the Subcommittee on Immigration, Citizenship, Refugees, Border Security and International Law on April 26, 2010.

H.R. 4952. The HAITI Act seeks to establish the Office of the Special Coordinator for Assistance to Haiti and to establish the Office of the Special Inspector General for Assistance to Haiti. Introduced and referred to the House Committee on Foreign Affairs on March 25, 2010.

H.R. 4961. The Haitian Private Sector Encouragement Act of 2010 intends to provide for the establishment of the Haitian-American Enterprise Fund. Introduced and referred to the House Committee on Foreign Affairs on March 25, 2010.

H.R. 5006. The White House Conference on Haiti Act of 2010 will require the President to call a White House Conference on Haiti. Introduced and referred to the Committee on Foreign Affairs as well as the Committee on Rules for consideration on April 13, 2010.

H.R. 5160. The Haiti Economic Lift Program Act of 2010 seeks to extend the Caribbean Basin Economic Recovery Act, to provide customs support services to Haiti, among other goals. Introduced and referred to the House Committee on Ways and Means on April 28, 2010. Passed House under suspension on May 5, 2010.

H.R. 5171. The Partnership with America's Rapid Rebuilding of Haiti Act of 2010 allows qualified and available United States construction workers and appropriate equipment to be sent to Haiti to assist Haitians in the rebuilding of their country after the devastating January 12, 2010, earthquake, as requested by the government of Haiti. Introduced and referred to the House Committee on Foreign Affairs on April 28, 2010.

H.Con.Res. 17. The resolution addresses the humanitarian assistance provided to Caribbean countries affected by past hurricanes and tropical storms. It acknowledges the affected countries' efforts to aid their citizens in recovery. The resolution also expresses support of the international assistance received by the Dominican Republic, Haiti, Jamaica, the Bahamas, Cuba and Turks and Caicos. Referred to the House Committee on Foreign Affairs January 9, 2009.

H.Con.Res. 165. Supports the yielding of temporary protected status for Haitian nationals who currently reside in the United States. Introduced July 17, 2009; referred to the Subcommittee on Immigration, Citizenship, Refugees, Border Security, and International Law on August 19, 2009.

H.Res. 1021. Expressing condolences to and solidarity with the people of Haiti in the aftermath of the devastating earthquake of January 12, 2010. Introduced January 20, 2010 and agreed to January 21, 2010.

H.Res. 1048. A resolution commending the efforts and honoring the work of the men and women of USNS Comfort and the United States Navy in the immediate response to those affected by the January earthquake. Introduced January 27, 2010 and agreed to in the House on February 23, 2010.

H.Res. 1066. A resolution recognizing the efforts of the U.S. Armed Forces, local first responders, and other members of Operation Unified Response for their swift and coordinated action in light of the devastation caused by the earthquake. Introduced on February 3, 2010 and agreed to on February 23, 2010 in the House.

H.Res. 1159. A resolution supporting efforts to address the crisis faced by Haitian orphans following the earthquake. Introduced and referred to the House Committee on Foreign Affairs on March 10, 2010.

H.Res. 1160. A resolution calling for the establishment of a Haiti Marshall Plan Committee to coordinate aid and development initiatives from multilateral development banks, international financial institutions, U.S. bilateral aid programs, and major international charities and nongovernmental organizations in response to the earthquake. Introduced and referred to the Committees on Financial Services and Foreign Affairs on March 10, 2010.

H.Res. 1277. A resolution commending the efforts and honoring the work of the State of Israel, its Defense Forces, and the Israeli people for their efforts to provide relief to Haiti in the aftermath of the earthquake in January, 2010. Introduced and referred to the House Committee on Foreign Affairs on April 20, 2010.

S. 2931. Would accelerate the income tax benefits for charitable cash contributions for the relief of victims of the earthquake in Haiti. Would treat cash contributions made after January 11, 2010, and before February 16, 2010, for the relief of earthquake victims in Haiti as having been made on December 31, 2009, for purposes of the tax deduction for charitable contributions. Introduced January 20, 2010, and read twice and referred to the Committee on Finance.

S. 2936. Haiti Assistance Income Tax Incentive Act. Would treat cash contributions made after January 11, 2010, and before March 1, 2010, for the relief of earthquake victims in Haiti as having been made on December 31, 2009, for purposes of the tax deduction for charitable contributions. Would deem a contribution as meeting the recordkeeping requirements of the Internal Revenue Code if the taxpayer produces a telephone bill showing the name of the donee organization and the date and amount of the contribution. Introduced January 20, 2010, read twice and referred to the Committee on Finance.

S. 2961. The Haiti Recovery Act would urge the Secretary of the Treasury to direct the U.S. Executive Director to each international financial institution to advocate: (1) the cancellation of all remaining debt obligations of Haiti, including obligations incurred after the date of the enactment of this Act and before February 1, 2012; (2) the provision of debt service relief for all of Haiti's remaining payments; and (3) that new assistance to Haiti should be primarily grants rather than loans.

Would urge the Secretary to: (1) instruct the U.S. Executive Director of the International Monetary Fund (IMF) to advocate the use of certain proceeds to provide debt stock relief, debt service relief, and grants for low-income countries that are eligible for the Poverty Reduction and Growth Facility or any other programs to assist low-income countries, including Haiti; (2) support the creation of an Inter-American Development Bank trust fund for Haiti that would leverage U.S. contributions and promote bilateral donations for the purpose of investing in Haiti's infrastructure; and (3) direct the U.S. Executive Director of the Inter-American Development Bank to increase earnings transfer to the Fund for Special Operations, which finances programming in Haiti and other weak economies in the Western Hemisphere.

Would urge the Secretary and the Secretary of State to use all appropriate diplomatic influence to secure cancellation of all remaining bilateral debt of Haiti. Introduced, read twice and referred to the Committee on Foreign Relations January 28, 2010. Placed on the Senate Legislative Calendar under General Orders, Calendar No. 276 and later held at the desk on March 9, 2010.

S. 2978. The Renewing Hope for Haitian Trade and Investment Act of 2010 would: direct the Secretary of Homeland Security (DHS), acting through the Commissioner for U.S. Customs and Border Protection (CBP), to commit sufficient CBP resources to: (1) enhance commercial assistance to promote trade among Haiti, the Dominican Republic, and the United States; (2) facilitate the preclearance of valid cargo from Haiti to the United States; (3) promote the efficient and secure movement of articles entering the United States under the Caribbean Basin Economic Recovery Act (CBERA); and (4) provide technical assistance and training to Haiti's customs service to improve production validation and compliance and understanding of U.S. customs procedures, such as the Electronic Visa Information System;

- amend CBERA to extend, in each succeeding one-year period through FY2013 (transition period), the duty-free treatment of certain imported knit apparel articles made in one or more Caribbean Basin Trade Partnership Act (CBTPA) beneficiary countries from yarns wholly formed in the United States;
- extend the value-added rule of origin for certain apparel and other textile articles imported from Haiti.
- extend, for the initial applicable one-year period, and each one-year period thereafter through FY2022, the duty-free treatment of apparel articles imported directly into the United States from Haiti or the Dominican Republic in amounts not to exceed specified percentages of the aggregate square meter equivalents of all apparel articles imported into the United States in the most recent 12-month period;
- extend, through December 20, 2013, the preferential treatment of wire harness automotive components manufactured in Haiti and imported into the United States, provided Haiti meets certain economic and political eligibility requirements;
- establish the Haiti Recovery and Investment Task Force to promote foreign investment in Haiti; direct the Comptroller General to report to Congress on the effectiveness of the trade preferences provided under the Haitian Hemispheric Opportunity Through Partnership Encouragement Act of 2006 (HOPE Act), as amended, as well as recommendations for improving such preferences. Introduced, read twice and referred to the Committee on Finance Feb. 2, 2010.

S. 2998. The Haitian Emergency Life Protection Act of 2010 would amend the Immigration and Nationality Act with respect to V- visa non-immigrants (spouses and minor children of lawful permanent residents who come to the United States to wait for completion of the immigrant visa process). This would include unmarried sons and daughters of such lawful permanent residents; Haitian nationals whose petition for a family-sponsored immigrant visa was approved on or before January 12, 2010; would authorize the Department of State to use secondary evidence to verify eligibility for such status or for immediate relative status; would sunset the provisions of this Act two years after its enactment. Introduced February 4, 2010, read twice and referred to the Committee on the Judiciary.

S. 3202. The Haitian-American Enterprise Fund Act would promote the strengthening of the Haitian private sector. Introduced and referred to the Senate Committee on Foreign Relations on April 14, 2010.

S. 3275. The Haiti Economic Lift Program Act of 2010 would extend the Caribbean Basin Economic Recovery Act, to provide customs support services to Haiti. Introduced, read twice, and referred to the Committee on Finance on April 28, 2010.

S.Con.Res. 61. A concurrent resolution expressing the sense of Congress that general aviation pilots and industry should be recognized for the contributions made in response to Haiti earthquake relief efforts. Introduced on April 27, 2010 and referred to the House Committee on Transportation and Infrastructure on May 3, 2010.

S.Res. 414. A resolution expressing the sense of the Senate on the recovery, rehabilitation, and rebuilding of Haiti following the humanitarian crisis caused by the January 12, 2010 earthquake in Haiti. Introduced February 9, 2010 and on February 24, 2010, placed on Senate Legislative Calendar under General Orders. Calendar No. 275. On March 4, 2010, the resolution was agreed to in Senate without amendment and with a preamble by unanimous consent.

APPENDIX A. EXPOSED POPULATION

USGS
science for a changing world

GSN USAID FROM THE AMERICAN PEOPLE

M 7.0, HAITI REGION

Origin Time: Tue 2010-01-12 21:53:10 UTC
Location: 18.46°N 72.53°W Depth: 13 km

PAGER
Version 7
Created: 1 day, 4 hours after earthquake

Estimated Population Exposed to Earthquake Shaking

ESTIMATED POPULATION EXPOSURE (k = x1000)		- -*	- -*	5,887k*	7,261k	1,049k	571k	314k	2,246k	332k
ESTIMATED MODIFIED MERCALLI INTENSITY		I	II-III	IV	V	VI	VII	VIII	IX	X+
PERCEIVED SHAKING		Not felt	Weak	Light	Moderate	Strong	Very Strong	Severe	Violent	Extreme
POTENTIAL DAMAGE	Resistant Structures	none	none	none	V. Light	Light	Moderate	Moderate/Heavy	Heavy	V. Heavy
	Vulnerable Structures	none	none	none	Light	Moderate	Moderate/Heavy	Heavy	V. Heavy	V. Heavy

*Estimated exposure only includes population within the map area.

Population Exposure

population per ~1 sq. km from Landscan

0	5	50	100	500	1000	5000	10000

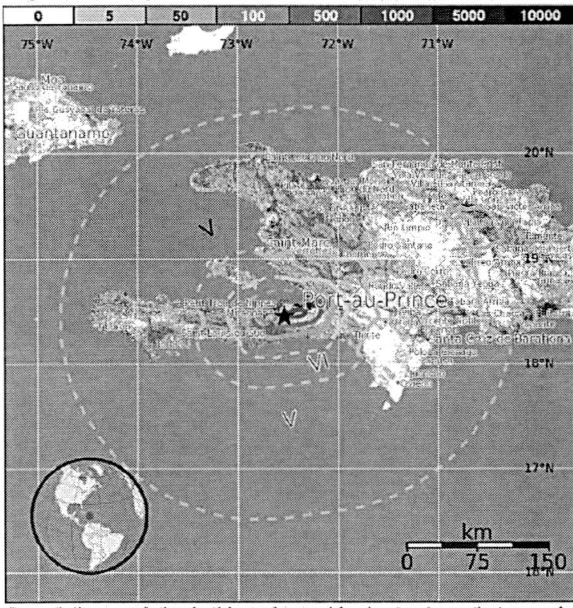

Selected City Exposure

MMI	City	Population
X	Grand Goave	5k
IX	Port-au-Prince	1,235k
IX	Carrefour	442k
IX	Petionville	108k
IX	Delmas 73	383k
IX	Croix des Bouquets	9k
VI	**Miragoane**	6k
V	**Verrettes**	49k
III	Santo Domingo	2,202k
III	**Guantanamo**	273k

bold cities appear on map (k = x1000)

Shaking Intensity MMI

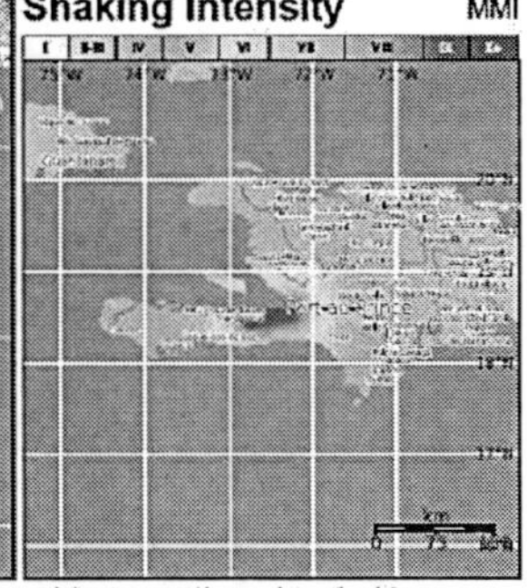

Overall, the population in this region resides in structures that are vulnerable to earthquake shaking, though some resistant structures exist. On June 24, 1984 (UTC), a magnitude 6.7 earthquake 329 km East of this one struck the Dominican Republic, with estimated population exposures of 320,000 at intensity VII and 2,964,000 at intensity VI, resulting in an estimated 5 fatalities. Recent earthquakes in this area have caused landslides that may have contributed to losses.

This information was automatically generated and has not been reviewed by a seismologist.

http://earthquake.usgs.gov/pager Event ID: us2010rja6

Source: U.S. Geological Survey, at http://earthquake.usgs.gov/earthquakes onepager.pdf.

Notes: The figure was generated by the Prompt Assessment of Global Earthquakes for Response (PAGER) system at the U.S. Geological Survey. PAGER is an automated system that rapidly assesses the number of people, cities, and regions exposed to severe shaking by an earthquake. Following the determination of earthquake location and magnitude, the PAGER system calculates the degree of ground shaking, estimates the number of people exposed to various levels of shaking, and produces a description of the vulnerability of the exposed population and infrastructure. This is version 7 of the PAGER output, accessed on January 14, 2010.

Figure A-1. An Estimate of the Population in Haiti and Surrounding Areas Exposed to Ground Shaking Caused by the January 12, 2010, Magnitude 7.0 Earthquake.

APPENDIX B. HAITI POPULATION MOVEMENT

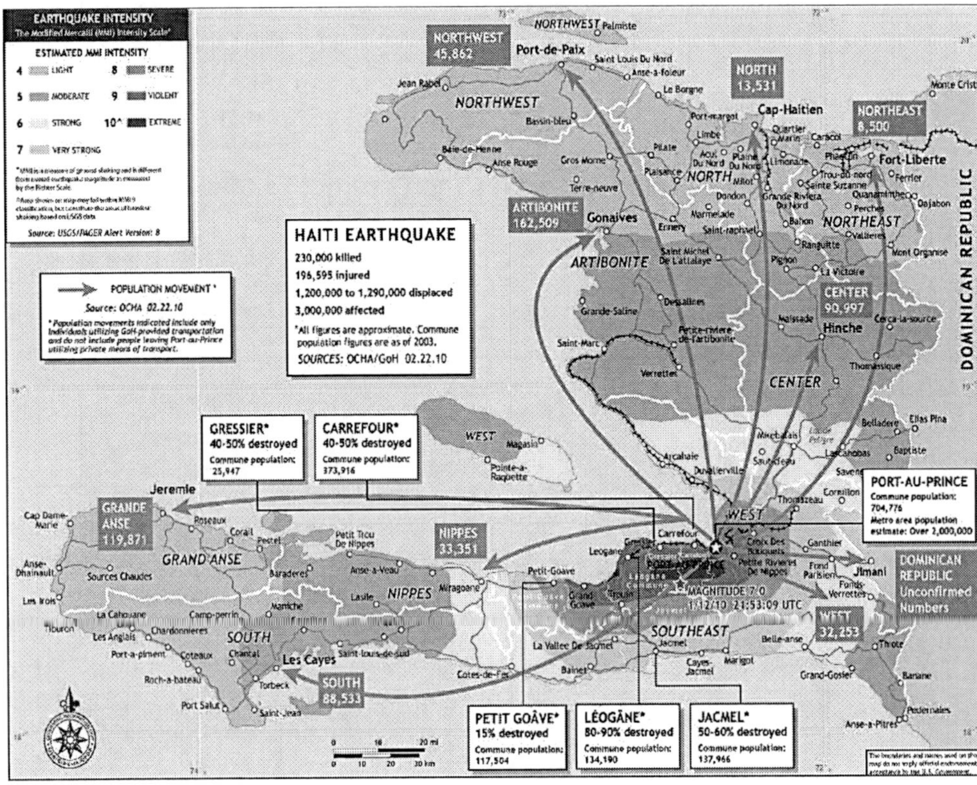

Figure B-1. Movement Out of Port-au-Prince.

APPENDIX C. EMERGENCY SHELTER GAP: APRIL 19, 2010

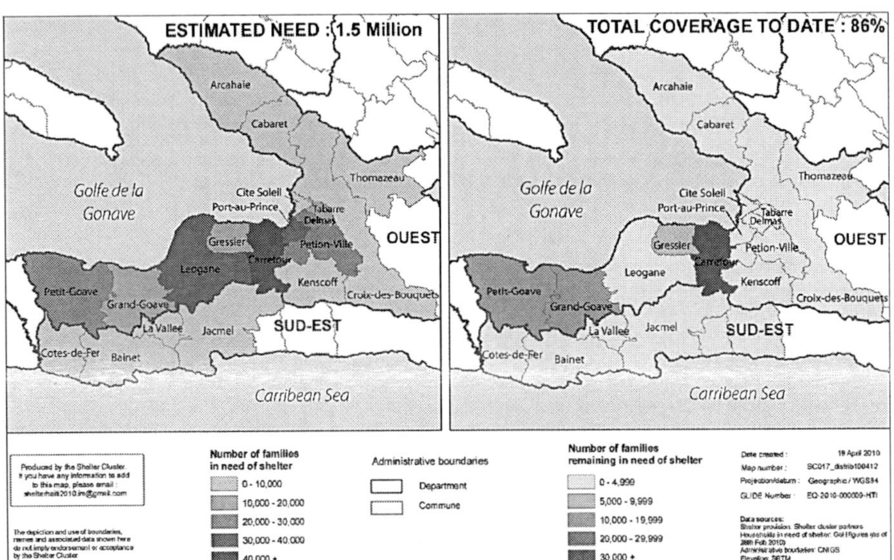

Figure C-1. Click and type title, or delete.

APPENDIX D. U.S. EARTHQUAKE ASSISTANCE TO HAITI

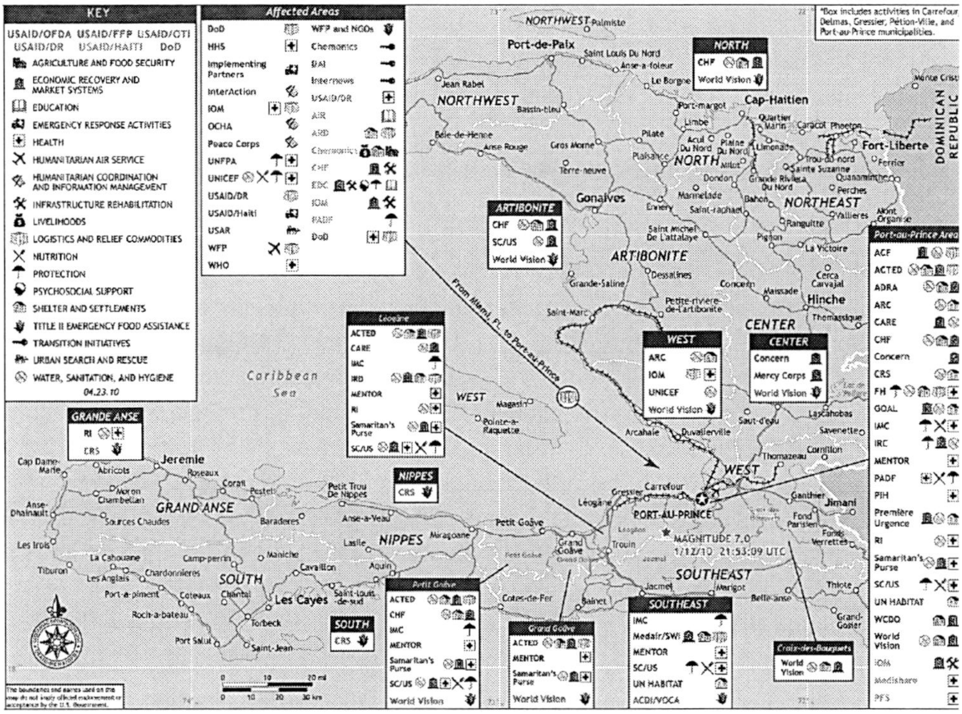

Figure D-1. USG Humanitarian Assistance.

APPENDIX E. THE U.S. GOVERNMENT EMERGENCY RESPONSE MECHANISM FOR INTERNATIONAL DISASTERS

The United States is generally a leader and major contributor to relief efforts in response to humanitarian disasters.[108] The President has broad authority to provide emergency assistance for foreign disasters and the U.S. government provides disaster assistance through several U.S. agencies. The very nature of humanitarian disasters—the need to respond quickly in order to save lives and provide relief—has resulted in a rather unrestricted definition of what this type of assistance consists of at both a policy and an operational level. While humanitarian assistance is assumed to provide for urgent food, shelter, and medical needs, the agencies within the U.S. government providing this support typically expand or contract the definition in response to circumstances. Funds may be used for U.S. agencies to deliver services or to provide grants to international organizations (IOs), international governmental and non-governmental organizations (NGOs), and private or religious voluntary organizations (PVOs). The U.S. Agency for International Development (USAID) is the U.S. government agency charged with coordinating U.S. government and private sector assistance. It also coordinates with international organizations, the governments of countries suffering disasters, and other governments.

The Office of Foreign Disaster Assistance (OFDA) in USAID's Bureau for Democracy, Conflict and Humanitarian Assistance (DCHA) provides immediate relief materials and personnel, many of whom are already abroad on mission. It is responsible for providing non-food humanitarian assistance and can quickly assemble Disaster Assistance Response Teams (DARTs) to assess conditions. OFDA has wide authority to borrow funds, equipment, and personnel from other parts of USAID and other federal agencies. USAID has two other offices that administer U.S. humanitarian aid: Food For Peace (FFP) and the Office of Transition Initiatives (OTI). USAID administers emergency food aid under FFP (Title II of P.L. 480) and provides relief and development food aid that does not have to be repaid. OTI provides post-disaster transition assistance, which includes mainly short-term peace and democratization projects with some attention to humanitarian elements but not emergency relief.

The Department of Defense (DOD) Overseas Humanitarian, Disaster and Civic Aid (OHDACA) funds three Dodd humanitarian programs: the Humanitarian Assistance Program (HAP), Humanitarian Mine Action (HMA) Program, and Foreign Disaster Relief and Emergency Response (FDR/ER). OHDACA provides humanitarian support to stabilize emergency situations and deals with a range of tasks including providing food, shelter and supplies, and medical evacuations. In addition the President has the authority to draw down defense equipment and direct military personnel to respond to disasters. The President may also use the Denton program to provide space-available transportation on military aircraft and ships to private donors who wish to transport humanitarian goods and equipment in response to a disaster.[109]

Generally, OFDA provides emergency assistance for 30 to 90 days after a disaster. The same is true for Department of Defense humanitarian assistance. After the initial emergency is over, assistance is provided through other channels, such as the regular country development programs of USAID.

The State Department also administers programs for humanitarian relief with a focus on refugees and the displaced. The Emergency Refugee and Migration Account (ERMA) is a contingency fund that provides wide latitude to the President in responding to refugee emergencies. Assistance to address emergencies lasting more than a year comes out of the regular Migration and Refugee Account (MRA) through the Population, Migration and Refugees (PRM) bureau. PRM assists refugees worldwide, conflict victims, and populations of concern to the United Nations High Commissioner for Refugees (UNHCR), often extended to include internally displaced people (IDPs). Humanitarian assistance includes a range of services from basic needs to community services.

Appendix F. Operation Unified Response: U.S. Military Units Participating

Major Commands

U.S. Southern Command http://www.southcom.mil/appssc/index.php
U.S. Air Mobility Command http://www.amc.af.mil
U.S. Air Force Special Operations Command http://www.afsoc.af.mil/
U.S. Navy Expeditionary Combat Command http://www.necc.navy.mil

Ground Units

82nd Airborne Division Brigade Combat Team http://www.bragg.army.mil/82DV/
22nd Marine Expeditionary Unit http://www.lejeune.usmc.mil/22ndMEU/

Naval Units

SS Cape May—Heavy-lift ship
SS Cornhusker State—Crane ship
USS Bataan—Amphibious assault ships
USS Fort McHenry, USS Carter Hall—Dock landing ships
USNS Comfort—Hospital ship
USNS Grasp—Salvage ship
USNS Sacagewea – Dry cargo ship
USNS Big Horn – Replenishment oiler
6 U.S. Coast Guard cutters

Appendix G. Donor Contributions and Pledges to Haiti in Response to the January 12, 2010, Earthquake

Country/Agency Donor	Monetary Pledge (USD)	In-kind Support Pledge
Afghanistan	$200,000	
Algeria	$1,000,000	
Andorra	$69,920	
Antigua and Barbuda	$37,037	
Argentina		medical equipment and relief supplies
Australia	$12,961,628	
Austria	$4,498,768	400 tents and first aid equipment
Azerbaijan	$499,978	
Bahamas	$25,000	
Bahrain	$1,000,000	disaster relief team
Bangladesh		medical team
Barbados		40-foot container filled with medical supplies, and two 20-foot containers with a variety of other items
Belgium	$1,873,376	search and rescue team, search dogs, medical team, field hospital, water purification system
Benin	$118,250	
Bolivia		food and blood
Bosnia & Herzegovina	$73,780	
Botswana	$128,100	
Brazil	$7,908,015	Air Force sent six flights with personnel, food and water, medications, emergency portable hospital. 50-member rescue team, rescue dogs
British Virgin Islands	$80,000	2 search and rescue officers
Bulgaria	$201,010	One and a half tons of humanitarian aid consisting of tents, blankets and bedding; 5-member medical team
Cambodia	$60,000	

Appendix G. (Continued).

Country/Agency	Monetary Pledge	
Canada	$81,331,088	search and rescue team, medical personnel, engineers, helicopters, supplies and equipment, two navy ships and 500 Canadian troops
Chile		15 metric tons of food and medicines, 61-member search-and rescue team
China	$10,539,457	60-member search-and-rescue team , 43-member medical staff, medicines and medical equipment, power generators, water purifying machines, 500 tents, clothing
Colombia	$50,000	231 personnel of search and rescue teams and health professionals; 21 trained dogs; 398 tons of humanitarian supplies and consignments of a military hospital, medication, medical supplies and equipment, food, water, water purifiers, sanitation supplies, vehicles, communication devices and search and rescue equipment
Congo, Democratic Republic of	$2,500,000	
Costa Rica		engineers, health care workers, doctors, and disaster experts
Croatia	$491,660	
Cuba		60 doctors added to 400 medical staff already in Haiti
Cyprus	$144,300	
Czech Republic	$1,134,358	
Denmark	$16,287,960	
Dominican Republic		food and water, rescue crews; 20 technicians to help re-establish telecommunications, 12 disaster management specialists, 46 doctors, 8 mobile clinics, 8 ambulances, 100 units of heavy construction equipment, transportation of humanitarian aid and injured victims; 28 mobile kitchens
Ecuador		search and rescue team, 5 tons of food
Egypt		medical supplies and personnel
El Salvador		12-member search and rescue team
Estonia	$356,421	
Finland	$8,654,958	two IT experts and IT module
France	$36,306,341	search and rescue teams, search dogs, three military transport helicopters, 130 tons of aid supplies, and troops, field hospital, medical personnel and medical supplies
Gabon	$1,000,000	
Georgia		15-member rescue team, 5 doctors, 40 tons of emergency supplies
Germany	$20,476,802	
Ghana	$3,000,000	
Greece	$928,115	team of physicians and healthcare professionals, rescue team, Hellenic Aid officials, pharmaceuticals; 50 tons of pharmaceutical supplies and food; transport members of the NGOs Doctors of the World, and Hellenic Rescue Team
Grenada	$100,000	
Guyana	$1,893,000	
Guatemala		rescue team
Hungary	$144,300	rescue team on the field (1 medical team of 4 doctors, 1 paramedic and 1 nurse plus team leader and approximately 1,5 tons of medicine and equipment
Iceland	$96,000	35-member search and rescue team
India	$5,000,000	

Country/Agency	Monetary Pledge	
Indonesia	$1,700,000	humanitarian workers (doctors, rescuers, electricity technicians, construction and telecommunication experts); tents, medicine, food, baby and children kits, one ambulance, one truck, and water purifying tool
Iran		30 tons of aid including food, tents, medicine
Ireland	$2,656,819	120 tons of emergency humanitarian aid and emergency supplies; Irish Aid Rapid Response Corps members
Israel		delegation which includes rescue forces, 40 doctors,24 nurses and IDF Medical Corps to set up a hospital
Italy	$9,603,946	interforce military contingent; Engineering Task Force consisting of 185 units equipped with: trucks, heavy tow trucks, containers, cranes, tank trucks, illuminated tower trailers, and a kitchen trailer; medical personnel of 37 doctors and nurses, 12 nurses; blankets, tents, beds
Jamaica		two Jamaica Defense Force vessels with technical, medical and military personnel and relief supplies
Japan	$25,327,154	team of 100 members (of whom about 40 are medical professionals), engineering unit, emergency relief goods
Jordan		12-bed military hospital; dispatched two planes carrying a mobile field hospital, rescue team, doctors and six tons of aid supplies that include food, medicine and clothing
Kazakhstan	$99,959	
Korea, Republic of	$1,200,000	relief goods; search and rescue teams and medical teams; 240 troops to Haiti to provide humanitarian assistance and rebuild areas in Leogane, west of the capital, Port-au-Prince
Kuwait	$1,000,000	100 tons of food, medical supplies, tents and blankets
Lebanon		25 tons of tents, 3 tons of medicine, vaccines and other supplies; sent aid workers to help in relief effort
Liberia	$50,000	
Liechtenstein	$144,661	
Lithuania	$268,554	
Luxembourg	$1,946,181	search and rescue team and dogs, six civil protection technical staff with localization material
Macedonia	$136,075	medical team
Malawi		150 metric tons of rice
Malta	$199,898	emergency assistance
Mauritius	$500,000	
Mexico		202-member rescue team, rescue equipment, 1,600 tons of emergency relief items
Moldova	$90,000	
Monaco	$144,000	
Mongolia	$20,000	
Montenegro	$68,700	
Morocco	$1,000,000	medical and pharmaceutical products
Netherlands	$5,050,504	60-member team with sniffer dogs
New Zealand	$1,432,193	
Nicaragua		special brigade to repair electric power lines; 31 military doctors of the Humanitarian Rescue Unit (URH) and humanitarian aid
Nigeria	$1,500,000	
Norway	$25,230,493	tents and water equipment, staff support to UN

Appendix G. (Continued).

Country/Agency	Monetary Pledge	
Panama		15 tons of food and other relief items, 10,000 bags of food, 45,000 pounds of food donated by the people of Panama, 21 rescue workers and 4 search dogs, 3 forensic doctors, 2 diplomats, 1 volunteer and 6 journalists; additional specialized medical team of 6 surgeons, 3 paramedics and 3 rescue workers to aid the medical relief efforts undergoing in Port-au-Prince
Paraguay		5,000 kilos of food; 400 blankets; 13 volunteers
Peru		54 tons of food and medicine
Philippines	$50,000	medical team
Poland	$2,031,169	54 rescuers and trained dogs, 4 tons of equipment; food and medical supplies, medicines (first aid items), blood and blood plasma, blankets, water
Portugal	$1,443,001	multidisciplinary team of 30 persons equipped to set a shelter camp for around 400 persons and provide medical care services, including small surgeries; provision of camping gear, tents, camp beds, medical supplies, generators, bed sheets, blankets, hygiene kits and kitchen supplies
Qatar		aircraft loaded with 50 tons of urgent relief materials; rescue team of 26 members and to set up field hospital
Romania	$72,150	
Russia	$5,700,000	100 tons of emergency relief, mobile air hospital, SAR team deployed, including medical personnel (psychological support)
Saint Lucia	$185,185	
Saint Vincent and the Grenadines	$100,000	
Country/Agency	Monetary Pledge	
Saudi Arabia	$50,000,000	
Senegal	$1,000,000	
Sierra Leone	$100,000	
Singapore	$50,000	
Slovakia	$324,675	field hospital, medical unit consists of 17 personnel/paramedics, 3 tons of medical/first aid items; 2 tons of relief materials of 15 tents, 58 sleeping bags, 58 beds and 58 blankets
Slovenia	$243,188	
South Africa		rescue teams of medical staff and engineers; 10 tons of search and rescue equipments and medical supplies
Spain	$34,516,159	A team of 40 experts in rescue operations and dogs; team of doctors and health specialists, medical equipment, humanitarian aid (tents, blankets, medical kits, water and sanitation material, etc.)
Suriname	$1,000,000	
Sweden	$22,481,493	
Switzerland	$11,663,710	dispatched an inter-disciplinary expert team (7 persons) to carry out a needs-assessment and a second team with 10 experts in the fields of water and sanitation, health and shelter; goods and transport contribution
Syrian Arab Republic		30 tons of humanitarian aid
Taiwan	$1,391,000	rescue team; medical team dispatched by MND; 83 tons of relief supplies; 450 tons of rice

Country/Agency	Monetary Pledge	
Thailand	$812,284	
Trinidad and Tobago	$1,000,000	relief aid
Tunisia	$1,000,000	
Turkey	$2,250,000	field hospital, 2 healthcare survey unit, medical unit (17 personnel/paramedics) and 10 tons of medical/first aid items, 20 tons of relief material consisting 200 tents, 2000 blankets, 145 set of kitchen materials, 1000 plastic bags for corps and 3 relief personnel, 1.5 tons of logistic equipment including food and cloths, 10 search and rescue team of AKUT Association
Turkmenistan	$500,000	
United Arab Emirates	$4,384,650	medical personnel, medical supplies, field hospital, water, transport contribution, food items, tents and blankets
United Kingdom	$33,007,214	64-strong UK search and rescue team with heavy lifting equipment
United States	$710,477,890	teams including up to 72 people, six search and rescue dogs and up to 48 tons of rescue equipment; USAID disaster experts who will help assess the situation in Haiti
Uruguay		rescue team of 5 military staff and 5 specialized canines
Venezuela		616 tons emergency relief, establishment of shipping and air corridor, medical and SAR teams; 116 tons of special machinery for reconstruction; In addition, the Venezuelan government provided food (10,000 tons), non-food items (30,000 tents), medicines and 225,000 barrels of diesel fuel and gasoline and transported SAR teams on six flights (total 107.5 tons; 520 aid personnel)
Vietnam	$50,000	
Allocation of funds from Red Cross / Red Crescent	$76,780,196	
Central Emergency Response Fund (CERF)	$37,976,462	
European Commission (ECHO)	$30,189,516	
Inter-American Development Bank	$200,000	
Pan American Development Foundation		shelter kits with tents, tarps, water purification tables; food; medical supplies; family tool kits, including shovels; telecommunications equipment; additional experts
World Bank	$100,000,000	

Source: OCHA at http://reliefweb.int/fts, Factiva and Lexis/Nexis.com news databases. Prepared by Julissa Gomez-Granger, Information Research Specialist, Knowledge Services Group.

Table G-1. Regional Donors to Haiti Relief Efforts

Country/Agency Donor	Monetary Pledge (USD)	In-kind Support Pledge
Antigua and Barbuda		
	$37,037	
Argentina		Medical equipment and supplies
Bahamas		
	$25,000	
Barbados		40-foot container filled with medical
		supplies, and two 20-foot containers with a variety of other items
Bolivia		Food and blood
Brazil		Air force sent six flights with
	$6,884,782 $10,000,000 (uncommitted pledge)	personnel, food and water, medications, emergency portable hospital, 50-member rescue team, rescue dogs; local purchase of food, by the Brazilian Embassy in Port-au-Prince, for school feeding rehabilitation programs in Haiti; assistance for rapid restoration of food production capacity
British Virgin Islands	$80,000	2 search and rescue officers to assist in response and relief
CARICOM		Team of 8 doctors and nurses specializing in anesthesiology, emergency care, general medicine, orthopedics, public health
Columbia	$50,000	7 cargo flights and one Navy ship with 231 personnel of search and rescue teams and health professionals: 21 trained dogs; 398 tons of humanitarian supplies and consignments of a military hospital, medication, medical supplies and equipment, food, water, water purifiers, sanitation supplies, vehicles, communication devices and search and rescue equipment
Costa Rica		Engineers, health care workers, doctors, and disaster experts
Cuba		60 doctors added to 400 medical staff already in Haiti
Dominican Republic		Food and water, rescue crews, 20 technicians to help re-establish telecommunications, 12 disaster management specialists, 46 doctors, 8 mobile clinics, 8 ambulances, 100 units of heavy construction equipment, transportation of humanitarian aid and injured victims, 28 mobile kitchens
Ecuador		Search and rescue team, 5 tons of food
El Salvador		12-member search and rescue team
Grenada		
	$100,000	
Guyana		
	$1,893,000	
Guatemala		Rescue team

Country/Agency Donor	Monetary Pledge (USD)	In-kind Support Pledge
Jamaica		Two Jamaica Defense Force vessels with technical, medical, and military personnel and relief supplies
Inter-American Development		
Bank	$200,000	
Mexico		202-member rescue team, rescue equipment, 1,600 tons of emergency relief items
Nicaragua		Special brigade to repair electric power lines; 31 military doctors of the Humanitarian Rescue Unit (URH) and humanitarian aid
Panama		15 tons of food and other relief items, 10,000 bags of food, 45,000 pounds of food donated by the people of Panama, 21 rescue workers and 4 search dogs, 3 forensic doctors, 2 diplomats, 1 volunteer and 6 journalists; additional specialized medical team of 6 surgeons, 3 paramedics and 3 rescue workers to aid the medical relief efforts undergoing in Port-au-Prince; medicines, gauze, bandages and plaster for making casts
Pan American Development Foundation		Shelter kits with tents, tarps, water purification tablets, food; medical supplies; family tool kits, including shovels; telecommunications equipment; additional experts
Paraguay		5,000 kilos of food; 400 blankets; 13 volunteers
Peru		54 tons of food and medicine
Saint Lucia		
	$185,185	
Saint Vincent and the Grenadines		
	$100,000	
Trinidad and Tobago		
	$1,000,000	
Uruguay		Rescue team of 5 military staff and 5 specialized canines
Venezuela		616 tons of emergency relief, establishment of shipping and air corridor, medical and SAR teams; 116 tons of special machinery for reconstruction; in addition, the Venezuelan government provided food, non-food items, medicines, and fuel and transported SAR teams on six flights (total 107.5 tons)

Source: OCHA at http://reliefweb.int/fts, Factiva and Lexis/Nexis.com news databases. Prepared by Julissa Gomez-Granger, Information Research Specialist, Knowledge Services Group.

Appendix H. How to Search for or Report on Individuals in Haiti[110]

Regarding U.S. Citizens in Haiti

The U.S. Embassy in Port Au Prince has set up a task force at the embassy that is taking calls as conditions permit. The embassy is working to identify U.S. citizens in Haiti who need urgent assistance and to identify sources of emergency help. U.S. citizens in Haiti are urged to contact the embassy via e-mail (ACSPaP@state) to request assistance. U.S. citizens in Haiti can call the embassy's Consular Task Force at 509-2229-8942, 509-2229-8089, 509-2229-8322, or 509- 2229-8672.

The Department of State has also created a task force to monitor the emergency. People in the United States or Canada with information or inquiries about U.S. citizens in Haiti may reach the Haiti Task Force at 888-407-4747, or by e-mail at Haiti-Earthquake@state. The Task Force phone number for those outside the United States and Canada is 202-501-4444.

In order to expedite requests for information about persons in Haiti, the following information is needed:

- full name
- date of birth
- citizenship
- time
- date
- place of last known location
- any contact information, such as a cell phone number or hotel/church number where the person could be reached
- the person's e-mail address
- passport information, if known

It is also important to provide the requestor's contact information, including phone numbers, relationship to the person about whom the inquiry is being made, and any special or emergency circumstances.

The following website provides information for calls regarding the welfare of American citizens in Haiti: http://haiti.usembassy.gov/service/emergency

The International Committee of the Red Cross also has a directory for missing and located persons in Haiti at http://www.familylinks.icrc.org/wfl/wfl_hti.nsf/bottin?openview.

Haitian Citizens in the U.S.

Haitian citizens in the U.S. trying to locate people in Haiti can register their names with the International Committee for the Red Cross at http://www.icrc.org/web/doc/siterfl0.nsf/htmlall/familylinks-haiti-eng?opendocument.

They can also call the Haitian Embassy in Washington, DC, at 202-332-4090, or the Haitian Consulate in New York City at 305-859-2003.

The *Miami Herald* provides a page to help families connect with family members at http://www.miamiherald.com/news/americas/haiti/connect/#vmix_media

Haitian citizens in the United States may also consult the directory on the International Committee of the Red Cross website for missing relatives, friends, and colleagues at http://www.familylinks.icrc.org/wfl/wfl_hti.nsf/bottin?openview.

U.S. Citizens with Pending Adoption Cases in Haiti

U.S. Department of Homeland Security Secretary Napolitano has announced a humanitarian parole policy for two categories of Haitian children in the process of being adopted by American citizens. This policy is explained at http://www.uscis.gov/portal/site/uscis/menuitem.5af9bb95919f35e66f614176543f6d1a/?vgnextoid= 9c22546ade146210VgnVCM100000082ca60aRCRD&vgnextchannel= 68439c7755cb9010VgnVCM10000045f3d6a1RCRD.

U.S. citizens with adoptions pending should send detailed information to the U.S. Department of Homeland Security (USDHS)/U.S. Citizenship and Immigration Services (USCIS) at HaitianAdoptions@dhs.gov. This e-mail address is intended only for submitting documents for pending adoption cases. Additional information may be found at the USCIS website at http://www.uscis.gov.

The Department of State has a dedicated website to Intercountry Adoptions at http://adoption.state. The Department of State also hosts a dedicated blog about Intercountry Adoptions at http://blogs. It has also established an e-mail address for questions at AskCI@state.gov.

The following information will need to be included in any inquiries addressed to either the Department of Homeland Security or the Department of State:

- Subject Line: "Haitian Adoption Information"
- Full name and contact information (including e-mail address) of parents
- Full name(s) of child(ren)
- Date(s) of birth of child(ren) (if possible)
- A brief summary of the status of the case
- Name and contact information for the orphanage

For more information on the U.S. government's response to Haiti's most vulnerable children, people may contact HaitianChildrenUSAID@usaid.gov or 202-712-0550.

Appendix I. How to Contribute to Relief Efforts[112]

How to Make Donations

According to Inter Action and other relief agencies, the best way to help is to donate financially to organizations responding to a disaster. Cash allows relief professionals to procure exactly what is needed in a disaster situation and ensure that donations are culturally, dietetically, and environmentally appropriate. Cash donations do not use up other scarce resources, such as transportation, staff time or warehouse space. As needed, cash can also be transferred quickly to where it is needed, helping bolster the economy of the disaster-stricken region.[113]

The White House suggests that those wishing to make a donation to relief efforts may contribute online through http://ClintonBushHaitiFund.org; text "QUAKE" to 20222 to charge a $10 donation to the Clinton Bush Haiti Fund (the donation will be added to your cell phone bill); or text "HAITI" to 90999, and $10 will be given automatically to the Red Cross, charged to your cell phone bill. Those wishing to donate may also visit InterAction, at http://www.interaction.org, to contribute to other non-governmental organizations.

The Department of State suggests that those who have significant in-kind contributions to make, such as a plane, a cargo ship, a team of doctors, portable generators, or large-scale water purification equipment, go to http://www.usaid.gov/helphaiti/.

USAID, through the non-profit organization, Aidmatrix Foundation, Inc., at http://www.aidmatrixnetwork.org/fema/PublicPortal/ListOfNeeds.aspx?PortalID=133, provides a searchable database to connect donors with needs. The lists of needs may be filtered by category, NGO, or item description.

A second option allows the donor to submit details of in-kind donations, and Aidmatrix will use the information to confirm the need with an NGO. Those who have significant in-kind contributions to make, such as a plane, a cargo ship, portable generators, or large-scale water purification equipment, may also access Aidmatrix at http://www.aidmatrixnetwork.org/fema/ PublicPortal/NewDonation.aspx?PortalID=133.

USAID recently released guidelines for doing business with USAID in Haiti. The following links provide information for grants, contracts, and unsolicited proposals:

- For contracting opportunities, FedBizzOpps, at http://www.fbo.gov.
- For grants opportunities, http://www.grants.gov.
- Unsolicited proposals guidelines, http://www.usaid.gov/business/business_opportunities/.

APPENDIX J. LINKS FOR FURTHER INFORMATION[114]

U.S. Government Agencies

U.S. Agency for International Development (USAID)
http://www.usaid.gov/helphaiti/
USAID AIDMATRIX: In Kind Donations
http://www.aidmatrixnetwork.org/fema/PublicPortal/ListOfNeeds.aspx?PortalID=133
USAID Humanitarian Assistance to Haiti for the Earthquake and Earthquake Affected Areas Maps
http://www.usaid.gov/helphaiti/documents/02.15.10-USAID-DCHAHaitiEarthquakeMapbook34.pdf

U.S. Department of Defense

http://www.defense.gov/home/features/2010/0110_haiti/
Major Military Support for Haiti at a Glance
http://www.defense.gov/home/features/2010/0110_haiti/military

U.S. Department of Health and Human Services

http://www.hhs.gov/haiti/
Centers for Disease Control Guidance for Relief Workers and Travelers to Haiti for Earthquake Response
http://wwwnc.cdc.gov/travel/content/news-announcements/relief

U.S. Department of Homeland Security
http://www.dhs.gov/journal/theblog/2010/02/guardians-report-in-from-haiti-video.html

U.S. Department of State
http://www.state

U.S. Department of State Embassy, Port-au-Prince
http://haiti.usembassy.gov/

U.S. Geological Survey
http://earthquake.usgs.gov/earthquakes/eqinthenews/2010/us2010rja6/

White House: Help for Haiti
http://www.whitehouse.gov/blog/2010/01/13/help-haiti

Information on the Haitian Earthquake

Haiti Earthquake Damage Map
http://www.reliefweb.int/rw/fullmaps_am.nsf/eca57e2740e7a919412569cf003180fa/0573522688593a18c12576aa00483368/$FILE/100112_07.45NYT_Haiti_Epicenter.pdf

Haiti Earthquake Intensity Map
http://www.reliefweb.int/rw/fullmaps_am.nsf/luFullMap/A4228B2905DCFFE6C12576A
B00285
81B/$File/map.pdf?OpenElement

Haiti Earthquake 2010, Multidisciplinary Center for Earthquake and Engineering Research (MCEER)
http://mceer.buffalo.edu/research/Reconnaissance/Haiti1-12-10/default.asp

Other Resources

Center for International Disaster Information (CIDI)
http://www.cidi.org/news/haiti-quake.htm

Earthquake Engineering Research Institute (EERI)
http://www.eqclearinghouse.org/20100112-haiti/

European Commission for Humanitarian Aid (ECHO)
http://ec.europa.eu/echo/aid/caribbean_pacific/haiti_earthquake_en.htm

InterAction
http://www.interaction.org/crisis-list/earthquake-haiti

Inter American Development Bank
http://www.iadb.org/haiti/index.cfm?lang=en&id=6407

International Monetary Fund
http://www.imf.org/external/country/hti/index.htm

Organization of American States: Pan American Disaster Foundation
http://www.panamericanrelief.org/

Pan American Health Organization
http://new.paho.org/disasters/index.php?option=com_content&task=view&id=1088&Itemid=1

Red Cross Movement

The American Red Cross
http://www.redcross.org

The International Federation of Red Cross and Red Crescent Societies
http://www.ifrc.org

The Haitian Red Cross
http://www.ifrc.org/address/ht.asp

The International Committee of the Red Cross
http://www.icrc.org/web/eng/siteeng0.nsf/htmlall/haiti

Relief Web
http://www.reliefweb.int/rw/dbc.nsf/doc108?OpenForm&emid=EQ-2010-000009-HTI&rc=2

United Nations

United Nations Children's Fund (UNICEF)
http://www.unicef.org/infobycountry/haiti_newsline.html

United Nations Habitat(UN-Habitat)
http://www.unhabitat.org/content.asp?cid=7780&catid=5&typeid=6&subMenuld=0

United Nations News Center
http://www.un.org/apps/news/infocusRel.asp?infocusID=91&Body=Haiti&Body1=

United Nations Office for the Coordination of Human Affairs (OCHA)
http://www.reliefweb.int/rw/RWFiles2010.nsf/FilesByRWDocUnidFilename/MUMA-84TW2F-full_report.pdf/$File/full_report.pdf

United Nations Stabilization Mission in Haiti (MINUSTAH)
http://www.un.org/en/peacekeeping/missions/minustah/

United Nations World Food Program (WFP)
http://www.wfp.org/stories/haiti-wfp-bring-food-devastating-quake

World Bank
http://web.worldbank.org/WBSITE/EXTERNAL/COUNTRIES/LACEXT/HAITIEXTN/0,,menuPK:338184~pagePK:141159~piPK:141110~theSitePK:338165,00.html

End Notes

[1] U.S. Geological Survey http://earthquake.usgs.gov/earthquakes January 13, 2010. The largest earthquake ever recorded was the 9.5 magnitude 1960 Chile earthquake, see http://earthquake.usgs.gov/earthquakes/world/historical_mag.php.

[2] United Nations, *Secretary-General, Briefing General Assembly on Haiti Disaster, Announces Release of $10 million in emergency Funds to Kick-Start Response*, SG/SM/12701; GA/10912, New York, NY, January 13, 2010.

[3] Eduardo A. Cavallo, Andrew Powell, and Oscar Becerra, *Estimating the direct economic damage of the Earthquake in Haiti*, Inter-American Development Bank, February 11, 2010, available at http://idbdocs.iadb.org/wsdocs/ getdocument.aspx?docnum=35072649.

[4] Mary Beth Sheridan, "Haiti earthquake damage estimated up to $14 billion," *Washington Post*, February 17, 2010.

[5] Government of Haiti, the United Nations, the European Union, the Inter-American Development Bank, and the World Bank, "Haiti Post Disaster Needs Assessment: draft executive summary," March 10, 2010. Available at http://www.haitisantodomingo2010.org/hsd2010/content/documents; and World Bank, *Haiti Damage and Need Assessment: World Bank Partners with Global Network of Scientists and Experts*, Press Release No. 2010/240/LCR, Washington, DC, January 26, 2010.

[6] United Nations Office for Least Developed Countries. Facts About Least Developed Countries (LDCs) available at http://www.unohrlls.org/UserFiles/File/Publications/Factsheet.pdf, accessed January 15, 2010.

[7] Bill Clinton, "How We Can Help Rebuild Haiti's Promise," *The Washington Post*, January 14, 2010.

[8] Cavallo et al., op. cit., pp. 11-12.

[9] Estimates of the death toll vary; most reports now indicate 200,000 or more dead.

[10] Initially, the government of Haiti facilitated the departure of several hundred thousand people from Port-au-Prince to outlying areas. An unknown number of individuals may have used private means to leave the city and seek shelter.

[11] Jacqueline Charles and Lydia Martin, "Without even a shirt, Rene Preval stays focused; President Rene Preval no longer has a palace or more than one borrowed shirt. But he survived the quake and is running his nation from a small room," *The Miami Herald*, January 20, 2010.

[12] Jacqueline Charles, Lesley Clark, and David Ovalle, et al., "Relief Efforts turn to long-term rebuilding," *The Miami Herald*, January 21, 2010.

[13] Government of Haiti, the United Nations, the European Union, the Inter-American Development Bank, and the World Bank, "Haiti Post Disaster Needs Assessment: draft executive summary," March 10, 2010. Available at http://www.haitisantodomingo2010.org/hsd2010/content/documents; and World Bank, *Haiti Damage and Need Assessment: World Bank Partners with Global Network of Scientists and Experts*, Press Release No. 2010/240/LCR, Washington, DC, January 26, 2010.

[14] Republic of Haiti, "Action Plan for the Reconstruction and National Development of Haiti," March 2010, available at http://www.haitisantodomingo2010.org/hsd2010/content/documents.

[15] Tim Padgett, "Haiti PM: We Can Rise Out of Our Postquake Squalor," *Time*, February 22, 2010. Haiti's "Action Plan for the Reconstruction and National Development of Haiti" is available at http://www.haitisantodomingo2010.org/ hsd2010/content/documents.

[16] University of Colorado at Boulder, "Industry Corruption, Shoddy Construction Likely Contributed to Haiti Quake Devastation," *E Science News*, January 14, 2010; Senator Loren Legarda, "Commentary: Earthquake Devastation Linked to Corruption," *Philippine Daily Inquirer*, January 25, 2010; "Will Endemic Corruption Suck Away Aid to Haitians?" *Reuters*, January 26, 2010.

[17] Prepared by Marjorie Ann Browne, CRS Specialist in International Relations, and Rhoda Margesson, CRS Specialist in International Humanitarian Policy, Foreign Affairs, Defense, and Trade Division.

[18] The Council resolution that created MINUSTAH is S/RES/1542 (2004). The resolution extending the mandate in October 2009 is S/RES/1892 (2009). The resolution adopted unanimously on January 19, 2010, is S/RES/1908 (2010).

[19] These figures are taken from the Secretary-General's April 2010 report to the Council, the police component consisted of 828 U.N. police and 1,350 formed police unit officers. Additional police from Bangladesh, Italy and India are expected in the coming weeks. See U.N. document S/2010/200, p. 9, para. 32.

[20] *Report of the Secretary-General on the United Nations Stabilization Mission in Haiti*, U.N. document [dated April 22, 2010] S/2010/200.

[21] S/2010/200, p. 14, para. 55. The legislative elections that had been scheduled for February 2010 had been postponed.

[22] S/2010/200, p. 14, para. 57.

[23] S/2010/200, p. 15, paras. 60-62. Ban's increase would "represent 200 United Nations police and three self-sustaining formed police units." According to the U.N. peacekeeping website, "A Formed Police Unit is a cohesive and trained team of 140 armed and well-equipped (for crowd control) police officers." See http://www.un.org/en/peacekeeping/ sites/police/policework.shtml.

[24] Mr. Mulet was Assistant Secretary-General, Office of Operations, in the Department of Peacekeeping Operations. He had preceded Mr. Annabi as SRSG for Haiti. On January 22, 2010, the United States and MINUSTAH had signed an agreement, clarifying their roles and responsibilities in coordinating international relief efforts with the government of Haiti.

[25] U.S. Department of State and U.S. Agency for International Development. FY2010 Haiti Supplemental Budget Justification. p. 44. This "peacekeeping" portion of the [total $143,489,000] request[ed] for the INCLE account would also pay for the construction of a temporary camp for the U.S. contingent and enough supplies to make them self-sustaining. U.S. assistance would also provide for the procurement of emergency supplies, replacement uniforms and equipment lost as a result of the earthquake, and purchase additional equipment needed to support an increased personnel contribution to MINUSTAH. Finally, the funds would support the training and equipping of a Formed Police Unit for the MINUSTAH mission in the areas of crowd/riot control and protection of U.N. facilities and personnel.

[26] Kim Bolduc, the U.N. Resident/Humanitarian Coordinator, was deployed to Haiti in November 2009.

[27] The head of each cluster (indicated in parentheses) reports to the Resident/Humanitarian Coordinator and works in partnership with all relevant actors in that particular sector. The clusters meet at least once daily.

[28] Information derived from a variety of sources, including USAID/OFDA, *Haiti-Earthquake*, Fact Sheet #13, FY2010 January 25, 2010; Pan American Health Organization, Emergency Operations Center Situation Report #12 Haiti Earthquake, January 25, 2010; OCHA, *Haiti Earthquake,* Situation Report #13, January 25, 2010.

[29] Many international actors have provided humanitarian relief to Haiti, either through financial contributions to the government of Haiti or aid organizations or by directly providing relief supplies and emergency personnel.

International recovery efforts are typically complex because they require coordination among numerous different actors, including other governments and international entities. Apart from U.N. agencies, those responding to humanitarian crises include international organizations, non-governmental organizations (NGOs), Private Voluntary Agencies (PVOs), and bilateral and multilateral donors.

30 The airport in the Dominican Republic has also been used as a humanitarian staging area to help with the coordination effort and allow for relief teams and supplies to get to Haiti by land through an established U.N. humanitarian coordinator.

31 Other agencies responding to the crisis include the Federal Emergency Management Agency (FEMA), Health and Human Services (HHS), and Department of Homeland Security (DHS).

32 See USAID website: http://www.usaid.gov/our_work/humanitarian_assistance/disaster template/index.html.

33 Prepared by Stephen Bowman, CRS Specialist in National Security, Foreign Affairs, Defense, and Trade Division.

34 United States Institute of Peace, Haiti: A Republic of NGOs? *Peacebrief* 23, April 26, 2010.

35 See, for example, Interaction, which is an alliance of U.S.-based international humanitarian and development NGOs at http://www.interaction.org.

36 Percentages total 101% due to rounding of request amounts.

37 As part of the United Nations' reform process, in March 2006, the CERF was launched based on several earlier resolutions approved by the U.N. General Assembly to strengthen the United Nations' capacity to respond to natural disasters and humanitarian emergencies. It is managed by the Emergency Relief Coordinator and head of UNOCHA. As an international, multilateral funding mechanism, the CERF aims to focus on early intervention, timely response, and increased capacity and support to underfunded crises. The funds come from voluntary contributions by member states and from the private sector. The CERF is seen by proponents as a way to enable the United Nations to respond more efficiently, effectively, and consistently to humanitarian crises worldwide. Others also believe that U.S. support for this idea is critical to sustaining momentum for donor contributions and continued support for the disaster relief fund.

38 See, for example, the Chronicle of Philanthropy at http://philanthropy.com/news/updates/index.php?id=10679 and http://philanthropy.com/news/updates/10632/donations

39 For more information, see http://clintonbushhaitifund.org.

40 Prepared by Maureen Taft-Morales, Specialist in Latin American Affairs, Foreign Affairs, Defense, and Trade Division.

41 The website for the conference is at http://www.haiticonference.org/. Much of the information for this section is drawn from documents at that site.

42 The Haiti Reconstruction Platform is available at http://www.refondation.ht/index.jsp?sid=1&id=1&pid=1.

43 Prepared by Martin Weiss, Specialist in International Trade and Finance, updated by Rebecca M. Nelson, Analyst in International Trade and Finance, Foreign Affairs, Defense, and Trade Division.

44 "World Bank to Provide an Additional $100 Million to Haiti, Following Earthquake," *World Bank*, January 13, 2010.

45 "World Bank Pledges Half Billion US Dollars for Haiti Over 14 Months," World Bank, March 31, 2010.

46 "December 2009—IDB Portfolio in Haiti, *Inter-American Development Bank*, December 17, 2009.

47 "Haiti earthquake: IDB redirects resources for emergency assistance and reconstruction," *Inter-American Development Bank*, January 13, 2010.

48 IDB approves historic expansion of capital, financial package for Haiti," *Inter-American Development Bank*, March 22, 2010.

49 Financing under the IMF's concessional lending facility, the Extended Credit Facility, carries a zero interest rate, with a grace period of 51/2 years, and a final maturity of 10 years. The Fund reviews the level of interest rates for all concessional facilities under the PRGT every two years.

50 Statement by IMF Managing Director Dominique Strauss-Kahn during the International Donors Conference Towards a New Future for Haiti, *International Monetary Fund*, March 31, 2010.

51 "France Asks Paris Club to Speed Up Haiti Debt Relief," *Reuters*, January 15, 2010.

52 "UPDATE 2-IADB governors commit to Haiti aid, fresh capital," *Reuters*, March 22, 2010.

53 World Bank Pledges Half Billion US Dollars for Haiti Over 14 Months, *World Bank*, March 31, 2010.

54 UN agency forgives debt owed by disaster-hit Haiti, *UN News*, April 23, 2010.

55 Heather Stewart , "Haiti Heads for Debt Crisis as Emergency Loans Pile Up after Earthquake," *The Observer*, January 31, 2010.

56 Pascal Fletcher, "Sarkozy Visits Haiti, Unveils Major Aid Package," *Reuters*, February 17, 2010. It is unclear why estimates of Haiti debt owed to France differ across news articles (€58 million vs. €56 million), but it is

reported that the debt relief provided by France to Haiti in mid-February allowed for cancellation of all of the debt owed by Haiti to France.

[57] "Taiwan Contemplating Haiti Debt Relief Decision," *Taiwan News*, January 21, 2010.

[58] IMF, "Statement by the IMF on Haiti," press release, February 24, 2010, http://www.imf.org/external/np/sec/pr/2010/pr1059.htm.

[59] Ibid.

[60] Caribbean Media Corporation, "Caribbean heads of state to visit Haiti to assess quake damage," *BBC Monitoring Americas*, January 14, 2010.

[61] Prepared by Maureen Taft-Morales, Specialist in Latin American Affairs, Foreign Affairs, Defense, and Trade Division.

[62] World Bank, *Country Report: Haiti*, 2010, http://go.worldbank.org/GBXI5JKM50.

[63] Ibid.

[64] "Haiti: UN Council Mission reports strides in security, worrisome poverty," States News Service, March 19, 2009.

[65] U.S. Dept. of State & U.S. AID, FY201ff0 Haiti Supplemental Budget Justification, p.4.

[66] "New Haitian Prime Minister-designate Profiled," BBC Monitoring Americas, November 5, 2009.

[67] World Bank, *Country Report: Haiti*, 2010, http://go.worldbank.org/GBXI5JKM50.

[68] "Rural Haitian Children Starving," *Associated Press*, November 21, 2008.

[69] Economist Intelligence Unit, "Country Report: Haiti ," November 2009, p. 12.

[70] Mike Blanchfield, "In recession, Haitians abroad send less money home," *Canwest News Service*, March 17, 2009.

[71] World Bank, "Income Distribution, Inequality, and Those Left Behind," *Global Economic Prospects 2007: Managing the Next Wave of Globalization*, p. 83. December 1, 2006.

[72] Republic of Haiti, *Growth And Poverty Reduction Strategy Paper*, November 2007, available at http://siteresources.worldbank.org/INTHAITIINFRENCH/Resources/HaitiDSNCRPEnglish.pdf.

[73] "New Haitian Prime Minister-designate Profiled," BBC Monitoring Americas, November 5, 2009.

[74] Le Rapport d'Evaluation des Besoins Apres Desastre Cyclones Fay, Gustav, Hanna et Ike (Haiti Post-Disaster Needs Assessment, PDNA), Prepared for the Government of the Republic of Haiti with the support of the World Bank, the United Nations System, and the European Commission, November 2008. Author's translation. Available in French at http://gfdrr.org/docs/Haiti_PDNA_2008.pdf. "Haiti: a New Paradigm" is available at http://idbdocs.iadb.org/WSDocs/ getDocument.aspx?DOCNUM=1942794.

[75] "Haiti Policy and Foreign Assistance Review," presentation for Congressional briefing, October 2009.

[76] "Haiti Policy and Foreign Assistance Review," presentation for Congressional briefing, October 2009.

[77] United Nations, *International Donors' Conference Towards a New future for Haiti*, Final Communique, New York, NY, 2010.

[78] Nina Shen Rastogi, "Was Haiti's ecosystem disrupted?," *The Washington Post*, February 9, 2010.

[79] Dan Schnitzer, "Avoid the Old Poverty Traps," *Foreign Policy*, January 19, 2010.

[80] Ibid.

[81] UNDP Update on Haiti Earthquake, January 20, 2010.

[82] Dan Schnitzer, "Avoid the Old Poverty Traps," *Foreign Policy*, January 19, 2010.

[83] For background information see CRS Report RL33769, *International Crises and Disasters: U.S. Humanitarian Assistance, Budget Trends, and Issues for Congress*, by Rhoda Margesson.

[84] Relief Web is a good source of information, although the accuracy is not guaranteed. See http://www.reliefweb.int. Obtaining an exact up-to-date record of all international contributions in response to an ongoing disaster is often not possible—in part because some assistance is not reported to governments or coordinating agencies—and in part because of the delay in their recording.

[85] Prepared by Ruth Ellen Wasem, Specialist in Immigration Policy, and Alison Siskin, Specialist in Immigration Policy, Domestic Social Policy Division. For further information, see CRS Report RS21349, *U.S. Immigration Policy on Haitian Migrants*, by Ruth Ellen Wasem.

[86] For additional information on Temporary Protected Status, see CRS Report RS20844, *Temporary Protected Status: Current Immigration Policy and Issues*, by Ruth Ellen Wasem and Karma Ester.

[87] U.S. Department of Homeland Security, " Secretary Napolitano Announces Humanitarian Parole Policy for Certain Haitian Orphans," fact sheet, January 18, 2010.

[88] Between January 18, 2010 and April 5, 2010, DHS authorized parole for more than 1,000 orphans under the special program, and approximately 340 cases were still being considered. U.S. Citizenship and Immigration

Services, "Special Humanitarian Parole Program for Haitian Orphans Draws to a Close at Request of Haitian Government," fact sheet, January 18, 2010.

[89] Some of the parents have admitted to relinquishing their children willingly in the belief that they would have a better life in the United States.

[90] The judge's name has also been spelled Saint-Vil in press reports. Joseph Guyler Delva, "Haiti Judge Rules for Release of U.S. Missionaries," *The Windsor Star*, February 11, 2010. Frank Bajak, "Lawyer Sees Ruling on Jailed Missionaries Thursday," *Associated Press*, February 11, 2010.

[91] "Missionary from Idaho Faces Haiti Trial," *Yahoo News,* April 27, 2010, http://news.yahoo.com/s/afp/20100427/ts_alt_afp/haitiquakeuscrimechildren_20100427114458.

[92] The other two activities included in $220 million request for HHS would cover costs associated with medical evacuations and costs for HHS public health activities in Haiti, which is discussed elswhere.

[93] Associated Press, "CG intercepts Haitian migrants, sends them home," *Navy Times*, February 16, 2010.

[94] Prepared by Sarah A. Lister, Specialist in Public Health and Epideb miology, Domestic Social Policy Division.

[95] Marc Caputo, "Florida Hospitals Insist They're Ready to Welcome Haitian Patients," *The Miami Herald*, January 31, 2010.

[96] HHS, "HHS Activates Additional Components of National Disaster Medical System to Help U.S. Hospitals Treat Survivors of Earthquake in Haiti," press release, February 1, 2010, http://www.hhs.gov/news.

[97] Communication from HHS Office of the Assistant Secretary for Preparedness and Response, March 8, 2010.

[98] HHS, "National Disaster Medical System Helping U.S. Hospitals Treat Survivors of Earthquake in Haiti," undated, http://www.hhs.gov/haiti/ndms_ushospitals.html.

[99] Team members are typically not federal employees, but are designated as intermittent federal employees during deployments.

[100] See CRS Report RL33053, *Federal Stafford Act Disaster Assistance: Presidential Declarations, Eligible Activities, and Funding*, by Keith Bea.

[101] Public Health Service Act § 2812; 42 U.S.C. § 300hh–11.

[102] HHS, "Payments for National Disaster Medical System (NDMS) Patients and Other Medical Evacuees from Haiti," questions and answers, March 19, 2010, provided to CRS by the HHS Office of the Assistant Secretary for Legislation.

[103] White House Office of Management and Budget, "Estimate #3—FY 2010 Emergency Supplemental Proposals in the FY 2011 Budget for Costs Associated with Relief and Reconstruction Support for Haiti following the Earthquake of January 12, 2010, ... ," March 24, 2010, pp. 13-14, http://www.whitehouse.gov/omb/assets/budget_amendments/ amendment_03_24_10.pdf.

[104] For further information, see CRS Report R41036, *Charitable Contributions for Haiti's Earthquake Victims*, by Molly F. Sherlock.

[105] Prepared by J. F. Hornbeck, Specialist in International Trade and Finance, Foreign Affairs, Defense, and Trade Division. For further information, see CRS Report RL34687, *The Haitian Economy and the HOPE Act*, by J. F. Hornbeck.

[106] For further information, see CRS Report RL34687, *The Haitian Economy and the HOPE Act*, by J. F. Hornbeck.

[107] Prepared by Maureen Taft-Morales, Specialist in Latin American Affairs, and Anne Leland, Information Research Specialist, Knowledge Services Group, Foreign Affairs, Defense, and Trade Division.

[108] For more information, see CRS Report RL33769, *International Crises and Disasters: U.S. Humanitarian Assistance, Budget Trends, and Issues for Congress*, by Rhoda Margesson.

[109] Section 402 of Title 10, named after former Senator Jeremiah Denton, authorizes shipment of privately donated humanitarian goods on U.S. military aircraft provided there is space and they are certified as appropriate for the disaster by USAID/OFDA. The goods can be bumped from the transport if other U.S. government aid must be transported.

[110] Prepared by Anne Leland, Information Research Specialist, Knowledge Services Group.

[111] Liz Heron, "Web Sites Offer Help for Reuniting Families Separated by Quake," *The Washington Post*, January 15, 2010, p. A9.

[112] Prepared by Anne Leland, Information Research Specialist, Knowledge Services Group.

[113] Inter Action at http://www.interaction.org/how-help.

[114] Prepared by Anne Leland, Information Research Specialist, Knowledge Services Group.

In: Haiti: Earthquake and Response
Editor: Rachael A. Donlon

ISBN: 978-1-61668-998-8

Chapter 2

CHARITABLE CONTRIBUTIONS FOR HAITI'S EARTHQUAKE VICTIMS*

Molly F. Sherlock

SUMMARY

On January 12, 2010, a magnitude 7.0 earthquake struck Haiti. As of January 25, 2010, the death toll was estimated to exceed 150,000. The earthquake and resulting aftershocks affected approximately 3 million people and caused significant damage to buildings and infrastructure. The earthquake has left an estimated 1 million Haitians homeless.

On January 22, 2010, President Obama signed into law the Haiti Assistance Income Tax Incentive Act (HAITI Act; P.L. 111-126). This legislation accelerates income tax benefits for charitable cash contributions for the relief of earthquake victims. Specifically, the legislation allows taxpayers making charitable contributions of cash made to organizations providing aid to earthquake victims after January 11, 2010, and before March 1, 2010, to take the associated charitable deduction on their 2009 income tax returns. The HAITI Act (H.R. 4462) was introduced in the House on January 19, 2010, and passed by a voice vote on January 20, 2010. The Senate had introduced companion legislation (S. 2936) on January 20, 2010, but passed the identical House legislation H.R. 4462 on January 21, 2010. Provisions in the HAITI Act are similar to those adopted under P.L. 109-1 following the 2004 Indian Ocean tsunami. The Joint Committee on Taxation (JCT) estimates that the HAITI Act would result in revenue losses of approximately $2 million over the 10-year budget window spanning FY2010 through FY2019.

Under current law, charitable contributions to 501(c)(3) charitable organizations from individuals, corporations, and estates and trusts are tax deductible in the year they are made. Individuals can deduct up to 50% of their adjusted gross income (AGI), phased-out for higher income individuals. Corporations can deduct up to 10% of their taxable income. Individuals and corporations can carry forward any unclaimed charitable deductions for up to five years. Total charitable giving in 2008 was $307.65 billion.

---

* This is an edited, reformatted and augmented version of a Congressional Research Services publication, datd January 26, 2010.

In the past, Congress has passed legislation to encourage charitable giving following natural disasters. Following the 2004 Indian Ocean tsunami, legislation was passed that allowed taxpayers making charitable contributions to aid tsunami victims in January 2005 to take the charitable deduction on their 2004 tax return. This provision is similar to the one enacted under the HAITI Act. In September 2005, following Hurricane Katrina, individual and corporate giving limits were suspended. The rules surrounding charitable contributions of food inventory and books were also relaxed to encourage in-kind giving.

The HAITI Act, like other tax policies, can be evaluated along the dimensions of efficiency and equity. Efficiency is greatest when the policy's marginal impact, the giving induced by the program, is large relative to the policy's inframarginal impact, the benefits given to those whose behavior was not directly caused by the tax policy. Using this framework, the HAITI Act is unlikely to be economically efficient. In general, tax benefits for charitable giving do not appear to significantly increase donations. Furthermore, tax deductions violate principles of vertical equity in that the benefits of tax deductions accrue disproportionately to higher income groups and provide larger benefits to those with a greater ability to pay.

Introduction

On January 12, 2010, a magnitude 7.0 earthquake struck Haiti. As of January 25, 2010, the death toll was estimated to exceed 150,000. The earthquake and resulting aftershocks affected approximately 3 million people and caused significant damage to buildings and infrastructure.[1] The earthquake has left an estimated 1 million Haitians homeless. In the United States, legislation has been enacted with the goal of promoting charitable donations for the earthquake victims in Haiti. Similar action was taken following the 2004 Indian Ocean tsunami and the 2005 Gulf Coast Hurricanes, when Congress enacted legislation to promote charitable giving to organizations providing aid to victims of these natural disasters.

On January 22, 2010, President Obama signed into law the Haiti Assistance Income Tax Incentive Act (HAITI Act; P.L. 111-126). This legislation accelerates income tax benefits for charitable cash contributions for the relief of earthquake victims. Specifically, the legislation allows taxpayers making charitable contributions of cash made to organizations providing aid to earthquake victims after January 11, 2010, and before March 1, 2010, to take the associated charitable deduction on their 2009 income tax returns. The HAITI Act (H.R. 4462) was introduced in the House on January 19, 2010, and passed by a voice vote on January 20, 2010. The Senate had introduced companion legislation (S. 2936) on January 20, 2010, but passed the identical House legislation H.R. 4462 on January 21, 2010. Provisions in the HAITI Act are similar to those adopted under P.L. 109-1 following the 2004 Indian Ocean tsunami. The Joint Committee on Taxation (JCT) estimates that the HAITI Act would result in revenue losses of approximately $2 million over the 10-year budget window spanning FY2010 through FY2019.

In the days following the earthquake in Haiti, many donors made $10 donations to earthquake victims via text message. The HAITI Act clarifies recordkeeping requirements for contributions by stating that a telephone bill satisfies the recordkeeping requirement so long as the bill states the name of the donee organization, the date of the contribution, and the amount of the contribution.

Senator Max Baucus, in a January 20, 2010, floor statement, stated "This is simple legislation that would make a big impact. It will make it a little easier for Americans to contribute to the victims of the Haiti disaster."[2] Senator Baucus went on to acknowledge that most Americans want to contribute in absence of this legislation, but noted that the HAITI Act would provide an additional incentive to give.

On January 20, 2010, Senators Charles E. Schumer and Kirsten Gillibrand introduced separate legislation in the Senate to encourage charitable donations to the earthquake victims of Haiti (S. 2937). This legislation, if enacted would suspend the individual and corporate charitable giving tax deduction limitations for cash donations directed toward Haitian earthquake victims. Similar provisions were enacted following Hurricane Katrina in 2005, discussed in further detail below. Additionally, S. 2937 would extend the enhanced deduction for food inventory, that was allowed to expire at the end of 2009. The enhanced deduction for food inventory provision is also part of Tax Extenders Act of 2009 (H.R. 4213), which passed the House on December 9, 2009. The Senate has not taken up the Tax Extenders Act of 2009.[3]

CURRENT LAW

Under current law, charitable contributions from individuals, corporations, and estates and trusts are tax deductible in the year they are made. Tax-exempt 501(c)(3) organizations, which include religious, educational, social welfare, health, international, and environmental organizations, are eligible to receive tax deductible charitable contributions. Contributions made to foreign agencies are generally not deductible.[4]

Individuals may claim deductions up to 50% of their adjusted gross income for qualified contributions to 501(c)(3) charitable organizations.[5] Itemized deductions are phased-out for high income individuals.[6] Individuals may carry forward unused charitable deductions for up to five years. In order to satisfy recordkeeping requirements, cash contributions can only be deducted if accompanied by a bank record (a cancelled check, bank statement, or credit card statement), a receipt, or payroll deduction documentation.

Corporations can deduct up to 10% of their taxable income for qualified charitable donations. Corporations can also carry forward unused charitable deductions for up to five years.

In 2008, individuals gave $229.28 billion. Corporations gave $14.5 billion, while bequests from estates totaled $22.66 billion in 2008. Grant making foundations gave $41.21 billion in 2008. Total charitable giving in 2008 was $307.65 billion.[7] Overall, giving between 2007 and 2008 fell by 5.7%, reflecting poor economic conditions.

Giving varies by the type of recipient organization. In 2008, religious organizations received 35% ($106.89 billion) of all gifts. Organizations focusing on human services received 9% ($25.88 billion) of all gifts, public society benefit organizations 8% ($23.88 billion), and organizations focusing on international affairs received 4% (13.30 billion) of all gifts in 2008.[8]

Table 1 provides the Joint Committee on Taxation's estimated tax revenue loss associated with the charitable contribution deduction for 2008 and 2009.[9] For 2008, tax expenditures for the charitable giving deduction were estimated to be $47 billion. Tax revenue loss from the charitable deduction is estimated to be $39 billion in 2009.[10] The reduction in estimated revenue loss is due to the weak economy.

Table 1. Tax Expenditure Estimates for the Charitable Giving Deduction billions of dollars

Year	Individual	Corporate	Total
2008	44.3	3.1	47.4
2009	36.2	3.0	39.2

Source: U.S. Congress, Joint Committee on Taxation, *Estimates of Federal Tax Expenditures for Fiscal Years 2008 - 2012*, committee print, prepared by Joint Committee on Taxation, 111th Cong., 1st sess., October 31, 2008, JCS-2-08 and U.S. Congress, Joint Committee on Taxation, *Estimates of Federal Tax Expenditures for Fiscal Years 2009 - 2013*, committee print, prepared by Joint Committee on Taxation, 111th Cong., 2nd sess., January 11, 2010, JCS-1-10.

NATURAL DISASTERS AND CHARITABLE CONTRIBUTIONS

The Indian Ocean Tsunami

Following the 2004 Indian Ocean tsunami, changes in the laws governing charitable giving were made in the act to accelerate the income tax benefits for charitable cash contributions for the relief of victims of the Indian Ocean tsunami (P.L. 109-1). The Indian Ocean tsunami struck on December 26, 2004. P.L. 109-1, enacted on January 7, 2005, allowed taxpayers that gave to nonprofit agencies aiding the victims of the tsunami in January 2005 to treat the donations as if they had been made during 2004, allowing such donations to be deducted by itemizing taxpayers on 2004 tax returns.[11] In the absence of this legislation, charitable deductions made in January 2005 would have been claimed on 2005 tax returns, filed in early 2006.

Under P.L. 109-1, only charitable contributions made in cash to charities providing aid to tsunami victims qualified for the accelerated deduction. Charitable contributions made in the form of property, capital assets, or in-kind contributions did not qualify for the accelerated deduction. Contributions made by credit card, check, or money order are generally considered cash contributions. The cash donation requirement was made to be consistent with requests made by organizations coordinating relief efforts that specifically requested monetary donations. Similar requests were made in January 2010 in response to the Haiti earthquake. Monetary donations reportedly reduce shipping costs associated with transporting relief items.[12]

Accelerating the deductibility of charitable contributions has quantifiable benefits. For example, assuming a marginal tax rate of 35%, a taxpayer who made a $1,000 contribution in January 2005 to a charity providing aid to tsunami victims, has a tax deduction for that gift worth $350 if claimed on the 2004 tax return. If the donor waits and claims the tax deduction on their 2005 return, assuming an interest rate of 5%, the present value of the future tax deduction is $333.[13] Hence, the taxpayer saved $17 by claiming the charitable deduction on their 2004 tax return.[14]

Hurricane Katrina

On September 23, 2005, President George W. Bush signed into law the Katrina Emergency Tax Relief Act of 2005 (KETRA; P.L. 109-73). Congress's intent when passing this legislation was to provide temporary tax relief to assist the victims of Hurricane Katrina. Included in this legislation were a number of provisions designed to encourage charitable giving.

Under KETRA, individual and corporate giving limits were temporarily suspended. Prior to KETRA, individuals could not claim charitable deductions that exceeded 50% of their AGI, while corporations' deductions were limited to 10% of their taxable base. Following KETRA, individuals could deduct cash contributions made between August 27, 2005, and January 1, 2006, up to 100% of their AGI. Further, charitable contributions were exempt from the Pease limitations.[15] For corporations, the deduction limit for cash contributions used for Hurricane relief was increased to 100%.

Provisions enacted under KETRA expanded charitable deductions available for inventory made between August 28, 2005, and January 1, 2006. Specifically, the enhanced deduction for food inventory, previously only available to C corporations, was extended and made available to other businesses, specifically S corporations and partnerships. C corporations were also allowed to take an enhanced deduction for donations of book inventory to schools.[16]

Additional provisions under KETRA increased the mileage rate deduction for individuals using personal vehicles for charitable purposes, excluded mileage reimbursements received from hurricane relief charities from a volunteer's gross income, and allowed individuals providing free housing to displaced Katrina victims to claim a personal exemption for such persons. The mileage rate deduction and mileage reimbursement provisions were temporarily enacted for relief efforts between August 24, 2005, and January 1, 2007.[17]

Table 2 presents estimates of the revenue losses associated with the charitable provisions enacted under KETRA. The charitable provision with the largest estimated revenue loss in KETRA was the suspension of individual and corporate giving limits, with a 10-year revenue loss estimate of $871 million. The total 10-year estimated revenue cost of the charitable provisions enacted under KETRA exceeds $1 billion.[18]

Charitable Giving in 2005

Relief efforts following the December 2004 Indian Ocean tsunami, the August and September 2005 Gulf Cost Hurricanes, and the October 2005 Pakistan earthquake were aided by U.S. based charitable donations. It is estimated that giving for natural disaster relief in 2005 was $7.37 billion (or 2.8% of total giving, which was $260.28 billion). Individuals were responsible for $5.83 billion (79%) of the gifts. Among the individual contributions, $4.25 billion (73%) went to hurricane relief, $1.54 billion (26%) went to tsunami relief, and $40 million (1%) went to Pakistan earthquake relief. Corporate donors gave $936 million in 2005 (13% of total disaster giving). Much of the corporate giving was in the form of in-kind donations and services. **Table 3** summarizes giving to aid the victims of natural disasters in 2005.

Table 2. Tax Expenditure Estimates for KETRA's Charitable Provisions millions of dollars

Provision	2006	2006-2010	2006-2015
Suspend individual and corporate giving limits for qualified contributions	819	753	871
Extension of the enhanced deduction for food inventory	20	20	20
Extension of the enhanced deduction for book inventory	5	5	5
Increase standard mileage rate for charitable use of a vehicle	17	29	29
Mileage reimbursement to charitable volunteers excluded from gross income	1	2	2
Additional personal exemption for those housing displaced individuals	96	128	128
Total	958	937	1,055

Source: CRS analysis of data provided in U.S. Congress, Joint Committee on Taxation, *General Explanation of Tax Legislation Enacted in the 109th Congress*, committee print, 109th Cong., January 17, 2007, JCS-1-00, pp. 789-90.

Table 3. 2005 Natural-Disaster-Related Giving billions of dollars

	Individuals	Corporations	Foundations	Total
Katrina	4.25	0.94	0.11	5.30
Tsunami	1.54	0.34	0.04	1.92
Earthquake	0.04	0.10	0.01	0.15
Total	5.83	1.38	0.16	7.37

Source: Giving USA Foundation, *Giving USA 2006: The Annual Report on Philanthropy for the Year 2005* (Indianapolis, IN: Giving USA Foundation, 2006), pp. 57-65.

Donations following natural disasters was also concentrated among certain charitable sectors. More than half of the giving for hurricane victims went to human services charities. For tsunami related giving, 36% of contributions went to charities serving the international community, while 31% went to human services charities.[19]

Aggregate measures reflect increased giving in 2005. Overall, inflation adjusted giving increased by 9.2% between 2004 and 2005 (see Figure 1). Between 2005 and 2006 overall giving fell by 2.4%. Giving as a percentage of GDP was 2.4% in 2005, 0.2 percentage points higher than the 2004 level of 2.2%. After 2005, giving as a percentage of GDP has remained below 2005 levels (see Figure 1).

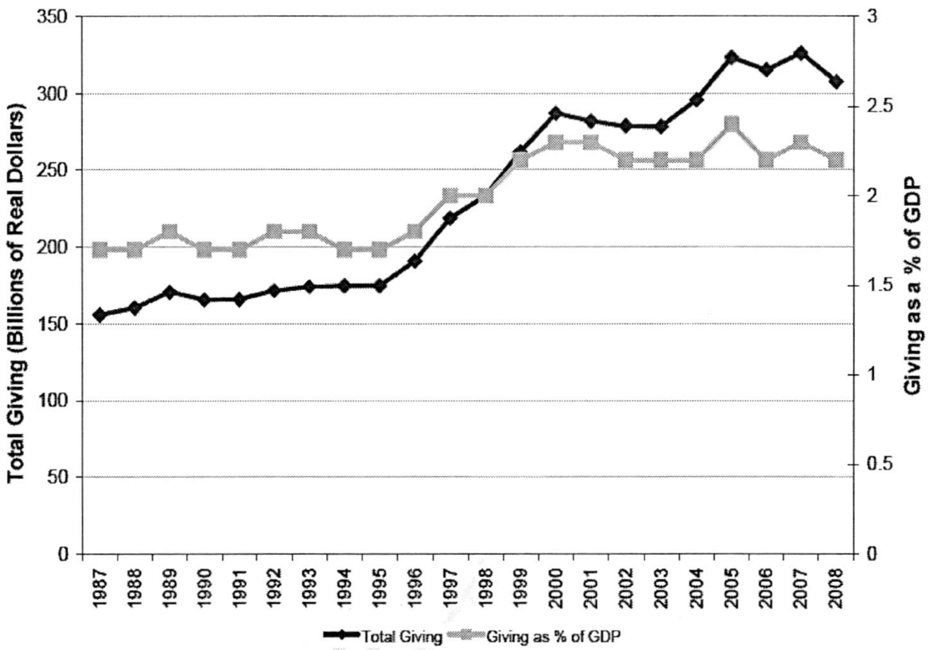

Source: CRS calculations based on data from Giving USA Foundation, Giving USA 2009: The Annual Report on Philanthropy for the Year 2008 (Indianapolis, IN: Giving USA Foundation, 2009).

ASSESSMENT OF THE HAITI ACT

The accelerated deduction provision of the HAITI Act, like deductions for charitable contributions generally, reduces the net cost of contributing, and is intended to encourage donations to disaster relief agencies that provide assistance to victims of the earthquake in Haiti. Like all tax benefits, the provision can be evaluated by assessing the impact on economic efficiency, equity, and complexity.

Economic Efficiency

The efficiency of the accelerated deduction depends on how much additional giving is induced by the provision, or the marginal impact of the program. The marginal impact of a tax break can be evaluated relative to the inframarginal impact. The inframarginal impact of a tax subsidy is the benefits given to individuals whose behavior is unchanged by the policy. In the case of the accelerated deduction, the inframarginal impact is the tax savings given to those who would have donated if the accelerated deduction had not been adopted. Efficient tax incentives are ones where the marginal impact is large relative to the inframarginal impact.

Since accelerated deductions reduce the price of charitable giving, empirical evidence on the price elasticity of charitable giving can provide insight regarding the expected marginal impacts. If charitable giving is price elastic, with a price elasticity greater than one, a $1 tax incentive would be expected to generate more than $1 in charitable giving.[20] If charitable giving is price inelastic, a $1 tax incentive would be expected to generate less than $1 in charitable giving. Empirical evidence suggests that the price elasticity for charitable giving is considerably less than one.[21] Since charitable contributions are relatively unresponsive to changes in the price of giving, the marginal impact of the accelerated deduction is likely to be small relative to the inframarginal impact. In general, theoretical models and empirical evidence suggest that tax benefits for charitable giving do not appear to significantly increase donations.[22]

There are also efficiency concerns surrounding the impact of accelerating the charitable deduction on the timing of giving. The provision is said to be efficient if it causes charitable giving to aid organizations to increase, without crowding-out donations to other charitable organizations. If charitable giving for disaster relief crowds-out other forms of charitable giving, altering the distribution of charitable giving across organizations or across time, the provision may not be efficient. Empirical evidence using individual-level data from the 2004 tsunami suggests that tsunami donations given in 2005 did crowd-out future charitable giving.[23] Other evidence, discussed in the context of donor fatigue below, suggests that donations for disaster relief did not contemporaneously crowd-out charitable giving to other organizations.

Evaluating the efficiency of accelerating the charitable donations deduction involves understanding both the timing and the nature of the contributions. From a timing perspective, the accelerated deduction is designed to encourage giving in early 2010, by allowing taxpayers to claim the charitable contributions on their 2009 tax returns.[24] However, to the extent that these donations were made without an expectation of an accelerated deduction provision, they are inframarginal. In the 13 days following the earthquake in Haiti, $470 million was donated to U.S. nonprofits providing aid to victims (this compares to $580 million in the eight days following Hurricane Katrina and $163 million in the nine days following the Indian Ocean tsunami).[25] Corporate giving in the 72 hours following the earthquake topped $43 million.[26] This immediate outpouring of support was not likely the result of expected changes to the tax code. Nonetheless, itemizers who made deductions during this time will benefit from the accelerated deduction provision if enacted.

Text Message Donations and the 2010 Earthquake in Haiti

Donations made via text message make up a significant portion of early giving to earthquake victims in Haiti. By Friday, January 22, 2010, the American Red Cross raised $29 million of their total $171 million through their text "HAITI" to 90999 campaign. Mobile phone users wishing to donate $10 to the American Red Cross can text "HAITI" to 90999. Their $10 donation is added to their mobile phone bill which can be used at tax filing time as a receipt.

Donors making small donations to the victims of Haiti's earthquake may be enticed to donate by convenience, rather than by the tax code. Many text-message donors may also be non-itemizers. Approximately 17% of all donations made to the American Red Cross were made via text message in the 13 days following the earthquake. Even for text message donors that do itemize, the value of the accelerated deduction is relatively small (17 cents for a taxpayer in the 35% bracket), and it seems unlikely that the prospect of such a donation would be driving behavior.

The "Hope for Haiti" celebrity telethon, broadcast on a number of major television networks, was also responsible for generating a large portion of the charitable contributions for Haiti relief efforts. The "Hope for Haiti" telethon, which aired on Friday, January 22, 2010, had raised $58 million by Saturday, January 23, 2010. A similar telethon following the 2004 tsunami raised $18.3 million.[27] It seems unlikely that most individuals donating through the telethon were doing so in response to tax incentives.

Congressional concern that hurricane-related giving may cause other forms of giving to fall (donor-fatigue) led to post-Katrina tax breaks for hurricane relief related charitable giving. The concept of donor fatigue rests on the notion that households set aside a fixed amount for charitable giving each year, and give until the funds are exhausted. Survey evidence following Hurricane Katrina suggests that donor fatigue did not cause donations to other charities to fall in 2005.[28] Given that disaster related giving was a relatively small part of overall giving in 2005 (less than 3%), it is unlikely that donor fatigue, if present, was large. Nonetheless, it is possible that donor fatigue was mitigated by Congressional action enhancing incentives for charitable contributions.

Equity

Provisions expanding the existing charitable deduction may raise equity concerns. First, only taxpayers who itemize deductions can take advantage of the accelerated deduction. Second, deductions violate vertical equity principles, as greater benefits accrue to higher income individuals.

The charitable contribution deduction can only be claimed by taxpayers who itemize deductions. In both 2006 and 2007, 35% of taxpayers itemized deductions.[29] Non-itemizing taxpayers claim the standard deduction ($5,700 for single filers in 2009). The logic behind the standard deduction is that it accounts for the tax-deductable activities of individuals choosing not to itemize deductions and simplifies tax compliance. As illustrated Table 4, the proportion of taxpayers who itemize increases for higher income taxpayers. The benefits associated with the charitable contribution deduction are also skewed towards higher income taxpayers. Just over 13% of taxpayers have an AGI between $100,000 and $200,000. Of these taxpayers, 81% itemize deductions. Nearly 25% of tax expenditures associated with the charitable deduction go to the $100,000 to $200,000 AGI group.

That charitable contributions violate vertical equity principles is a consequence of a progressive tax system. The value of a deduction to a taxpayer is based on their marginal tax rate. Take, for example, two taxpayers both donating $2,000 to victims of natural disasters. One taxpayer is in the 10% income tax bracket, while the other is in the 35% income tax bracket. Assuming both taxpayers itemize deductions, the value of the deduction to the first taxpayer is $200 while the value of the deduction to the second taxpayer is $700. The taxpayer with the higher income in the 35% tax bracket receives a larger tax benefit from the deduction than the lower income taxpayer.

Table 4. Distribution of Taxpayers, Itemizers, and Charitable Tax Deduction Benefits by Income 2008

Adjusted Gross Income (AGI) Group	% of All Returns in AGI Group	% in AGI Group that Itemize	% Charitable Tax Expenditure Benefits Received by AGI Group
Below $10,000	17.1%	2.5%	0.0%
$10,000 - $20,000	13.4%	5.7%	0.2%
$20,000 - $30,000	10.1%	12.7%	0.8%
$30,000 - $40,000	9.2%	21.5%	1.7%
$40,000 - $50,000	8.3%	32.9%	2.5%
$50,000 - $75,000	15.0%	45.2%	8.8%
$75,000 - $100,000	10.0%	59.8%	9.1%
$100,000 -$200,000	13.2%	80.9%	24.8%
Above $200,000	3.7%	94.0%	52.1%

Source: CRS calculations using data from U.S. Congress, Joint Committee on Taxation, *Estimates of Federal Tax Expenditures for Fiscal Years 2009 - 2013*, committee print, prepared by Joint Committee on Taxation, 111th Cong., 2nd sess., January 11, 2010, JCS-1-10.

Complexity

Accelerating charitable deductions adds complexity to the tax code and increases administrative burdens. Taxpayers have the option of claiming earthquake related charitable contributions on either their 2009 or 2010 income taxes. Since the 2009 forms have been printed and mailed to taxpayers, some taxpayers may not be aware of this change or understand the option to accelerate deductions.

It is also possible that accelerating charitable contributions increases the monitoring burden on the Internal Revenue Service. Since taxpayers would be allowed to claim earthquake related charitable contributions in either 2009 or 2010, there may be an increased monitoring burden on the IRS to ensure that deductions are only claimed in one year, and not both. In practice, it is likely that the tax revenue loss associated with taxpayers accidentally or intentionally claiming contributions in both 2009 and 2010 is less than the cost to the IRS of monitoring contributions.

Appendix. Giving Following Natural Disasters

Figure A-1 presents trends in charitable giving in the days following recent natural disasters. The immediate charitable giving response to aid the victims of the earthquake in Haiti was larger than the charitable giving response following the 2004 Indian Ocean tsunami or Hurricane Katrina in 2005. In the five days following the earthquake in Haiti, charitable donations were estimated to be $150 million. This compares to $108 million in the five days following Hurricane Katrina and $30 million in the four days following the tsunami. Charitable donations for the victims of the earthquake in Haiti have continued to outpace donations following the tsunami but now lag behind the pace of donations following Hurricane Katrina.

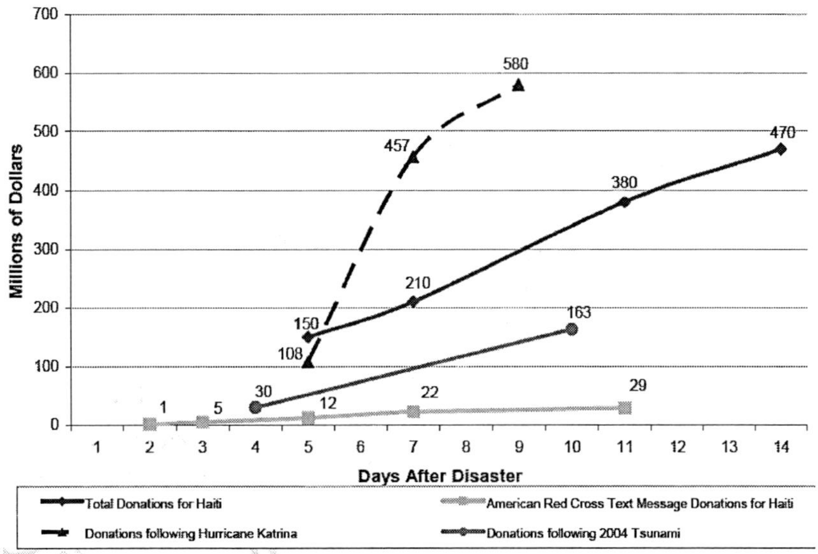

Source: CRS analysis using data reported by the Chronicle of Philanthropy, http://www.philanthropy.com

Figure A-1. Donations in the Days Following Natural Disasters.

End Notes

1 Internal Revenue Service, "IRS Announces Qualified Disaster Treatment for Haiti," IR-2010-11, January 22, 2010.

2 Committee on Finance, "Floor Statement of Senator Max Baucus (D-Mont.) Regarding Assistance Income Tax Incentive Act," press release, January 20, 2010, http://finance.senate.gov/press/Bpress/2010 press/prb012010b.pdf.

3 JCT estimates suggest that extending the enhanced deduction for charitable contributions food inventory through 2010 would result in $78 million in revenue losses. See U.S. Congress, Joint Committee on Taxation, *Estimated Revenue Effects of H.R. 4213, the "Tax Extenders Act of 2009"*, committee print, 111th Cong., 1st sess., December 7, 2009, JCX-59-09.

4 Generally, charitable donations made to foreign charitable organizations cannot be deducted on U.S. income tax returns (tax treaties with Mexico, Canada, and Israel allow charitable contributions to Mexican, Canadian, and Israeli charities to be tax deductible under some circumstances). Deductions that are made to a U.S. organization that transfers funds to a foreign charitable organization are deductible so long as the U.S. organization maintains control over the use of the funds. For more information see http://www.irs.gov/publications/p54/ch05.html#en_US_publink100047565.

5 Gifts of appreciated property are subject to a 30% limit. Gifts to foundations are further limited.

6 In practice, the itemized deduction phase-out works as an income tax surcharge. Specifically, Δ(tax liability) = t * 0.01*(AGI – Threshold), where t is the marginal tax rate and the threshold is set annually under Pease. For details on the limitations on itemized deductions see CRS Report R40508, *Personal Exemption Phaseout (PEP) and Limitation on Itemized Deductions (Pease)*, by Maxim Shvedov, p.9.

[7] Giving USA Foundation, *Giving USA 2009: The Annual Report on Philanthropy for the Year 2008* (Indianapolis, IN: Giving USA Foundation, 2009), p. 210.

[8] CRS Report R40919, *An Overview of the Nonprofit and Charitable Sector*, by Molly F. Sherlock and Jane G. Gravelle.

[9] Tax revenue losses attributable to tax provisions that allow for special exclusions, exemptions, or deductions from income or provisions that provide special tax credits, preferential tax rates, of defer tax liability are called tax expenditures.

[10] The Joint Committee on Taxation had previously estimated 2009 charitable giving tax expenditures for individuals to result revenue losses of $46.2 billion and $3.2 billion of revenue losses for corporations, for total revenue losses of $49.4 billion. The downward revision reflects the deterioration of the U.S. economy, where both income and asset values declined. See U.S. Congress, Joint Committee on Taxation, *Estimates of Federal Tax Expenditures for Fiscal Years 2008 - 2012*, committee print, prepared by Joint Committee on Taxation, 111th Cong., 1st sess., October 31, 2008, JCS-2-08.

[11] This measure was enacted quickly. After being proposed on January 4, 2005, the measure was introduced in both the House and the Senate on January 6, 2005, and signed into law by the President on January 7, 2005.

[12] Kurt Ritterpusch, "Congressional OK Expected for Bill to Extend 2004 Tsunami Relief Deductions," *The Daily Tax Report*, January 7, 2005, p. G1.

[13] The present value, or value today, of a future payment is calculated by multiplying the value today by $[1/(1+r)]$, where r is the interest rate.

[14] Givers do have the potential to reduce this loss in the absence of an accelerated deduction provision by adjusting their withholdings. Taxpayers that make large donations can reduce withholding estimated tax payments at the time of the donation rather than waiting until their taxes are filed to receive the benefit.

[15] For a discussion of the Pease limitations see CRS Report R40508, *Personal Exemption Phaseout (PEP) and Limitation on Itemized Deductions (Pease)*, by Maxim Shvedov.

[16] Both provisions, now part of the "tax extenders," were extended through 2007 by the Pension Protection Act of 2006 (P.L. 109-280) and through 2009 by the Emergency Economic Stabilization Act of 2008 (EESA; P.L. 110-343). Both provisions are now expired. If enacted, the Tax Extenders Act of 2009 (H.R. 4213) would extend these provisions through 2010. For more information see CRS Report RL32367, *Certain Temporary Tax Provisions Expiring in 2009 ("Extenders")*, by Pamela J. Jackson and Jennifer Teefy and CRS Report RL34608, *Tax Issues Relating to Charitable Contributions and Organizations*, by Jane G. Gravelle and Molly F. Sherlock.

[17] For further information see CRS Report R40434, *Charitable Standard Mileage Rate: Considerations for the 111th Congress*, by Nonna A. Noto.

[18] A number of the charitable provisions enacted under KETRA were expanded under the Gulf Opportunity Zone Act of 2005 (P.L. 109-135) to apply to charitable contributions made to qualified organizations providing aid to victims of Hurricanes Rita and Wilma.

[19] Giving USA Foundation, *Giving USA 2006: The Annual Report on Philanthropy for the Year 2005* (Indianapolis, IN: Giving USA Foundation, 2006).

[20] The price elasticity of charitable giving is the percentage change in giving divided by the percentage change in the price of giving, where the price of giving is $(1-t)$, and t is the marginal tax rate. While the price elasticity is negative, it is quoted in absolute value.

[21] For a review of the literature see CRS Report R40518, *Charitable Contributions: The Itemized Deduction Cap and Other FY2010 Budget Options*, by Jane G. Gravelle and Donald J. Marples.

[22] Thomas L. Hungerford, "Tax Expenditures: Good, Bad, or Ugly?," *Tax Notes*, October 23, 2006, p. 329.

[23] Sarah Brown, Mark N. Harris, and Karl Taylor, *Modeling Charitable Donations to an Unexpected Natural Disaster: Evidence from the U.S. Panel Study of Income Dynamics*, Institute for the Study of Labor (IZA), IZA DP No. 4424, Bonn, Germany, September 2009.

[24] The Appendix illustrates the amount of giving in the days following the earthquake in Haiti and compares the level of giving in Haiti to the levels of giving following Hurricane Katrina and the 2004 tsunami.

[25] Caroline Preston and Nicole Wallace, "Donations to Aid Haiti Exceed $470-Million, Chronicle Tally Finds," *The Chronicle of Philanthropy*, January 25, 2010.

[26] Chris Herring and Dana Mattioli, "Companies Send Aid to Haiti," *The Wall Street Journal*, January 19, 2010.

[27] Caroline Preston, "Haiti Telethon Raises More Than $58-Million," *The Chronicle of Philanthropy*, January 25, 2010.

[28] The Conference Board, "Consumer Confidence Survey," April, 2006.

[29] Internal Revenue Service (IRS), Statistics of Income (SOI), Individual Income Tax Returns Preliminary Data, 2007. Available at http://www.irs.gov/pub/irs-soi/09sprintaxreturn.pdf.

In: Haiti: Earthquake and Response
Editor: Rachael A. Donlon

ISBN: 978-1-61668-998-8

Chapter 3

U.S. Immigration Policy on Haitian Migrants*

Ruth Ellen Wasem

Summary

The environmental, social, and political conditions in Haiti have long prompted congressional interest in U.S. policy on Haitian migrants, particularly those attempting to reach the United States by boat. While some observers assert that such arrivals by Haitians are a breach in border security, others maintain that these Haitians are asylum seekers following a decades old practice of Haitians coming by boat without legal immigration documents. Migrant interdiction and mandatory detention are key components of U.S. policy toward Haitian migrants, but human rights advocates express concern that Haitians are not afforded the same treatment as other asylum seekers.

The devastation caused last year by the January 12, 2010, earthquake in Haiti led Department of Homeland Security (DHS) Secretary Janet Napolitano to grant Temporary Protected Status (TPS) to Haitians in the United States at the time of the earthquake. The scale of current humanitarian crisis—estimated thousands of Haitians dead and reported total collapse of the infrastructure in the capital city of Port au Prince—resulted in this TPS announcement on January 15, 2010.

Secretary Napolitano gave humanitarian parole to Haitian children who were legally confirmed as orphans eligible for intercountry adoption by the government of Haiti and who were in the process of being adopted by U.S. residents prior to the earthquake. P.L. 111-293, the Help HAITI Act of 2010, authorizes the DHS Secretary to adjust to legal permanent residence (LPR) status those Haitian orphans who were granted parole from January 18, 2010, through April 15, 2010.

Those Haitians who are deemed Cuban-Haitian Entrants are among the subset of foreign nationals who are eligible for federal benefits and cash assistance. Those Haitians who are newly arriving legal permanent residents, however, are barred from the major federal benefits and cash assistance for the first five years after entry. The Supplemental Appropriations Act, 2010 (H.R. 4899, P.L. 111-212), includes funding to cover additional costs for federal benefits and cash assistance resulting from Haitian evacuees.

---

* This is an edited, reformatted and augmented version of a Congressional Research Services Publication, dated January 21, 2011.

According to the U.S. Department of State (DOS), there were 54,716 Haitians who had approved petitions to immigrate to the United States at the time of the earthquake and who were waiting for visas to become available. Advocates for Haitians continue to request that Secretary Napolitano give humanitarian parole to those Haitians with approved petitions for visas. Proponents of expediting the admission of Haitians with family in the United States maintain that it would relieve at least some of the humanitarian burden in Haiti and would increase the remittances sent back to Haiti to provide critical help as the nation tries to rebuild. Those opposed to expediting the admission of Haitians assert that it would not be in the national interest, nor would it be fair to other foreign nationals waiting to reunite with their families.

More broadly, there are concerns that the crisis conditions in Haiti—most recently, the outbreak of cholera and the unexpected return of deposed dictator Jean-Claude "Baby Doc" Duvalier— may trigger mass migration from the island. Agencies within DHS that are the leads in handling a potential mass migration include the U.S. Coast Guard (interdiction); Customs and Border Protection (apprehensions and inspections); Immigration and Customs Enforcement (detention and removal); and the U.S. Citizenship and Immigration Services (credible fear determinations). The balancing of DHS's border security and immigration control responsibilities during an ongoing humanitarian crisis poses a challenge.

CURRENT CONTEXT

The devastation caused by the January 12, 2010, earthquake in Haiti focused world attention on the humanitarian crisis and prompted U.S. leaders to reconsider policies on Haitian migration.1 Most recently, the U.S. Centers for Disease Control (CDC) has stated: "An epidemic cholera strain has been confirmed in Haiti, causing the first cholera outbreak in Haiti in at least 100 years."2 Some members of Congress have long criticized the interdiction and mandatory detention of Haitians who attempted to enter the United States without proper immigration documents as too harsh given country conditions. Proponents of immigration control policies have held sway for many years in large part because they argue that more lenient treatment of Haitians would serve as a magnet for illegal migration from the poorest nation in this hemisphere. Whether the balance should tip more toward humanitarian policies as a consequence of the humanitarian crises that resulted from last year's earthquake in Haiti is an issue before the 112th Congress.

IMMIGRATION TRENDS

Migration by Sea

The phenomenon of Haitians coming to the United States by boat without proper travel documents dates back at least to the 1970s. An estimated 25,000 Haitians were among the mass migration of over 150,000 asylum seekers who arrived in South Florida in 1980 during the Mariel boatlift.3 The U.S. Coast Guard, as described below, has been interdicting vessels carrying Haitians since 1981. **Figure 1** presents the U.S. Coast Guard data on Haitian migrants that the Coast Guard has encountered on boats and rafts in the years following the Mariel boatlift. Most notably, there was a drop of migrants after the Haitian elections in 1990 followed by a dramatic upturn after the 1991 coup (discussed in "Crisis After the Coup" below). As country conditions in Haiti4 and U.S. policy responses to the surges in Haitian boat people are considered, the spikes and valleys in Figure 1 become more understandable. Since FY1998, the Coast Guard had interdicted over 1,000 Haitians each year, with 1,198 in FY2006 and 1,610 in FY2007. Haitian interdictions were second only to Cuban interdictions (2,868) in FY2007. In FY2009, interdictions of 1,782 Haitians led all other countries. As of August 17, 2010, the Coast Guard had interdicted 1,271 Haitians in FY2010.5

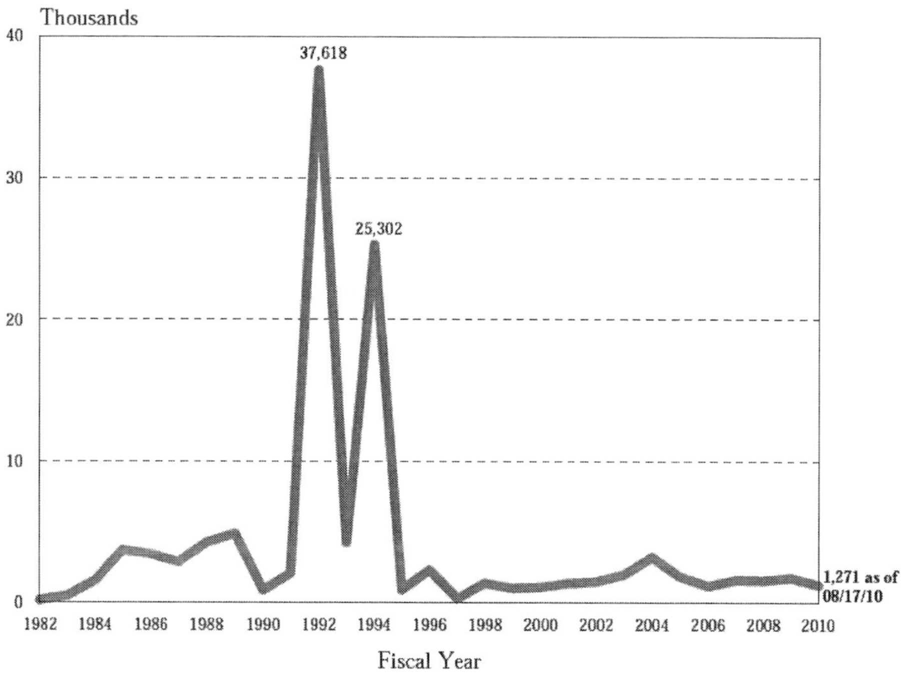

Source: CRS presentation of U.S. Coast Guard data.

Figure 1.U.S. Coast Guard Interdiction of Haitians, 1982-2010.

Not all Haitian migrants are interdicted by the Coast Guard, as witnessed in the widely televised landing of over 200 Haitians in Biscayne Bay, FL, in October 2002. Another noteworthy incident occurred in December 2001 when a boat bringing 167 Haitians ran aground in South Florida. In March 2007, the U.S. Border Patrol apprehended 100 Haitians who came ashore near Miami. During 2007, there were also reports of deaths at sea when boats with Haitians capsized or—in one report—caught fire.[6]

Haitians Currently Residing in the United States

In the 2000 Decennial census of the U.S. population, there were 532,000 persons reporting Haitian ancestry residing in the United States, 185,000 who were born in the United States and another 154,000 who were naturalized U.S. citizens. According to Congressional Research Service estimates based on the 2006-2008 American Community Survey (a 1% sample of the U.S. population conducted by the Bureau of the Census), there were approximately 757,000 persons reporting Haitian ancestry in the United States (margin of error is 15,233). Of these, an estimated 523,000 were born in Haiti and migrated to the United States.[7] These data do not indicate the immigration status of the Haitians, and a portion of these Haitians may have become naturalized U.S. citizens.[8]

During the period 2001-2008, there were 166,281 Haitians who became legal permanent residents (LPRs) in the United States, according to the Office of Immigration Statistics (OIS) in the Department of Homeland Security (DHS).[9] OIS estimates that approximately 230,000 Haitians were LPRs as of 2008. [10] Many of these Haitians adjusted to LPR status as a result of the Haitian Refugee Immigration Fairness Act of 1998, which is discussed more fully below.

There are currently no reliable estimates of Haitians residing in the United States without authorization (i.e., unauthorized aliens). The OIS estimates on unauthorized alien residents in 2008 do not include Haiti among the top 10 sending countries. [11] DHS has indicated that 35,110 Haitians have submitted petitions for Temporary Protected Status, as discussed below.[12]

POLICY EVOLUTION

Post-Mariel Policy

The Carter Administration labeled Haitians as well as Cubans who had come to the United States during the 1980 Mariel Boatlift as "Cuban-Haitian Entrants" and used the discretionary authority of the Attorney General (e.g., humanitarian parole) to admit them. It appeared that the vast majority of Haitians who arrived in South Florida did not qualify for asylum according to the newly enacted individualized definition of persecution in §207-208 of the Immigration and Nationality Act (INA, as amended by the Refugee Act of 1980).[13] Subsequently, an adjustment of status provision was included in the Immigration Reform and Control Act (IRCA) of 1986 that enabled Cuban-Haitian Entrants to become legal permanent residents (LPRs).[14]

Interdiction Agreement

In 1981, the Reagan Administration reacted to the mass migration of asylum seekers who arrived in boats from Haiti by establishing a program to interdict (i.e., stop and search certain vessels suspected of transporting undocumented Haitians). This agreement, made with then-dictator Jean-Claude Duvalier, authorized the U.S. Coast Guard to board and inspect private Haitian vessels on the high seas and to interrogate the passengers. At that time, the United States generally viewed Haitian boat people as economic migrants deserting one of the poorest countries in the world.

Under the original agreement, an inspector from the former Immigration and Naturalization Service (INS) and a Coast Guard official, working together, would check the immigration status of the passengers and return those passengers deemed to be undocumented Haitians. An alien in question must have volunteered information to the Coast Guard or INS inspector that she or he would be persecuted if returned to Haiti in order for the interdicted

Haitian to be considered for asylum. Ultimately, INS would determine the immigration status of the alien in question. From 1981 through 1990, 22,940 Haitians were interdicted at sea. Of this number, INS considered 11 Haitians qualified to apply for asylum in the United States.

Crisis after the Coup

The 1991 military *coup d'etat* deposing Haiti's first democratically elected President, Jean Bertrand Aristide, however, challenged the assumption that all Haitian boat people were economic migrants. The State Department reportedly hesitated on whether the Haitians should be forced to return given the strong condemnation of the *coup* by the United States and the Organization of American States. By November 11, 1991, approximately 450 Haitians were being held on Coast Guard cutters while the Administration of then-President George H. W. Bush considered the options. The former Bush Administration lobbied for a regional solution to the outflow of Haitian boat people, and the United Nations High Commissioner for Refugees (UNHCR) arranged for several countries in the region—Belize, Honduras, Trinidad and Tobago, and Venezuela—to temporarily provide a safe haven for Haitians interdicted by the Coast Guard. Some of the other countries in the region were each willing to provide safe haven for only several hundred Haitians. Meanwhile, the Coast Guard cutters were becoming severely overcrowded, and on November 18, 1991, the United States forcibly returned 538 Haitians to Haiti.

Pre-Screening and Repatriation

The options for safe havens in third countries in the region proved inadequate for the sheer numbers of Haitians fleeing their country, and the George H. W. Bush Administration began treating the Haitians fleeing by boat as asylum seekers. The Coast Guard took them to the U.S. naval base in Guantanamo, Cuba, where they were pre-screened for asylum in the United States. During this period, there were approximately 10,490 Haitians who were paroled into the United States after a pre-screening interview at Guantanamo determined that they had a credible fear of persecution if returned to Haiti. On May 24, 1992, citing the surge of Haitians that month, then-President Bush ordered the Coast Guard to intercept all Haitians in boats and immediately return them without interviews to determine whether they were at risk of persecution. The Administration offered those repatriated the option of in-country refugee processing.[15]

Safe Haven and Refugee Processing

The repatriation policy continued for two years, until then-President Bill Clinton announced that interdicted Haitians would be taken to a location in the region where they would be processed as potential refugees. The refugee processing policy lasted only a few weeks—June 15 to July 5, 1994. Much like the George H. W. Bush Administration, the Clinton Administration cited the exodus of Haitian boat people as a reason for suspending refugee processing. Instead, the new policy became one of regional "safe havens" where interdicted Haitians who expressed a fear of persecution could stay, but they would not be allowed to come to the United States. In 1993, in-country refugee processing was further expanded to Les Cayes and Cape Haiten. In December 1997, President Clinton instructed the Attorney General to grant deferred enforced departure (DED) to Haitians for one year. Currently interdicted Haitians who expressed a fear of persecution are taken for a credible fear hearing at the Guantanamo Bay detention center. If deemed a refugee, they are resettled in the third country. In 2005, only nine of the 1,850 interdicted Haitians received a credible fear hearing and, of those, one man was granted refugee status.[16]

Haitian Refugee Immigration Fairness Act (HRIFA)

When Congress enacted the Nicaraguan Adjustment and Central American Relief Act (NACARA) in November 1997 that enabled Nicaraguans and Cubans to become legal permanent residents and permitted certain unsuccessful Central American and East European asylum applicants to seek another form of immigration relief, it opted not to include Haitian asylum seekers. The following year, Congress enacted the Haitian Refugee Immigration Fairness Act (HRIFA) of 1998 (S. 1504/H.R. 3049) that enabled Haitians who filed asylum claims or who were paroled into the United States before December 31, 1995, to adjust to legal permanent residence. HRIFA was added to the FY1999 Omnibus Consolidated and Emergency Supplemental Appropriations Act (P.L. 105-277) at the close of the 105th Congress.[17] P.L. 110-161 deleted the requirement that the Comptroller General of the United States submit to Congress a status report on HRIFA applications every six months.

Mandatory Detention of Aliens in Expedited Removal

Since enactment of the Illegal Immigration Reform and Immigrant Responsibility Act (IIRIRA) of 1996 (P.L. 104-208), aliens arriving in the United States without proper immigration documents are immediately placed in expedited removal. If an alien expresses a fear of being forced to return home, the immigration inspector refers the alien to a asylum officer who determines whether the person has a "credible fear." IIRIRA requires that those aliens must be kept in detention while their "credible fear" cases are pending.[18] As a result, those Haitians who do make it to U.S. shores and do express a fear of repatriation are placed in detention. After the credible fear determination, the case is referred to an Executive Office for Immigration Review (EOIR) immigration judge for an asylum and removal hearing (during which there is no statutory requirement that aliens be detained).

EOIR granted asylum to 570 Haitians and denied asylum to 2,522 Haitians in FY2006.[19] In FY2007, EOIR granted asylum to 587 Haitians, and 510 Haitians were granted asylum in FY2008 (the most recent year for which data are available), an approval rate of 4.6% and 4.8%, respectively.[20] With respect to the number of Haitians in detention, according to Immigration and Customs Enforcement, as of January 19, 2010, there were 488 Haitians in detention, most of whom (415) were criminal aliens.[21]

Procedural Practices and Controversies

National Security

The former INS published a notice clarifying that certain aliens arriving by sea who are not admitted or paroled are to be placed in expedited removal proceedings and detained (subject to humanitarian parole) in November 2002.[22] This notice concluded that illegal mass migration by sea threatened national security because it diverts the Coast Guard and other resources from their homeland security duties. Then-Attorney General John Ashcroft expanded on this rationale in his April 17, 2003, ruling that instructs EOIR immigration judges to consider "national security interests implicated by the encouragement of further unlawful mass migrations" in making bond determinations regarding release from detention of unauthorized migrants who arrive in "the United States by sea seeking to evade inspection."[23] The case involved a Haitian who had come ashore in Biscayne Bay, FL, on October 29, 2002, and had been released on bond by an immigration judge. EOIR's Board of Immigration Appeals (BIA) had upheld his release, but the Attorney General vacated the BIA decision.[24]

Parole from Detention

In 2002, DOJ acknowledged that it instructed field operations "to adjust parole criteria with respect to all inadmissible Haitians arriving in South Florida after December 3, 2001, and that none of them should be paroled without the approval of headquarters."[25] The Administration of President George W. Bush maintained that paroling Haitians (as is typically done for aliens who meet the credible fear threshold) would encourage other Haitians to embark on the "risky sea travel" and "potentially trigger a mass asylum from Haiti to the United States." The Bush Administration further argued that all migrants who arrive by sea posed a risk to national security and warned that terrorists may pose as Haitian asylum seekers. Critics of the Bush Administration's Haitian parole policy focused on the 167 Haitians detained after their boat ran aground in South Florida on December 3, 2001, a majority of whom reportedly passed the initial credible fear hearing. Critics maintained that the Haitians were being singled out for more restrictive treatment.[26] They challenged the view that Haitians posed a risk to national security and asserted that the term was being construed too broadly, being applied arbitrarily to Haitians, and wasting limited resources.[27] OIS has reported that Haitians made up 2% of the 378,582 foreign nationals detained by DHS Immigration and Customs Enforcement in 2008.

Access to Legal Counsel

Concern also arose that the detention of Haitians interferes with access to legal counsel to aid with their asylum cases. According to congressional testimony, attorneys in South Florida for the detained Haitians maintained that they face various obstacles, including restricted hours to meet with clients and a serious lack of adequate visitation space. Pro bono lawyers working with Haitians argued that they experienced long delays waiting to see clients.[28] Others pointed out that the expedited removal provisions in INA were enacted to do just that—*expedite removals*. Aliens without proper immigration documents who try to enter the United States, they argued, should not be afforded the same procedural and legal rights as aliens who enter legally.

TEMPORARY PROTECTED STATUS[29]

The issue of Haitian TPS has arisen several times in the past few years, most notably after the U.S. Ambassador declared Haiti a disaster in September 2004 due to the magnitude of the effects of Tropical Storm Jeanne. More recently, a series of tropical cyclones in 2008 resulted in hundreds of deaths and led some to label the city of Gonaives uninhabitable.[30] The Administration of President George W. Bush did not grant TPS or other forms of blanket relief to Haitians, nor was legislation that would have provided TPS to Haitians, such as H.R. 522 in the 110th Congress, enacted. Opponents of Haitian TPS traditionally argue that it would result in an immigration amnesty for unauthorized Haitians and foster illegal migration from the island.

The scale of the current humanitarian crisis—estimated thousands of Haitians dead and reported total collapse of the infrastructure in the capital city of Port au Prince—led DHS to announce on January 13, 2010, that it is temporarily halting the deportation of Haitians. "TPS is in the range of considerations we consider in a disaster," stated DHS Deputy Press Secretary Matthew Chandler, "but our focus remains on saving lives."[31] In the 111th Congress, Representative Alcee Hastings has introduced H.R. 144, which would make nationals from Haiti eligible for TPS status. H.R. 264, introduced by Representative Shelia Jackson-Lee, includes a sense of Congress that "the Secretary of Homeland Security should be more liberal with respect to Haiti in deciding whether to designate that country for temporary protected status."

On January 15, 2010, DHS Secretary Napolitano granted TPS for 18 months to Haitian nationals who were in the United States as of January 12, 2010. She stated: "Providing a temporary refuge for Haitian nationals who are currently in the United States and whose personal safety would be endangered by returning to Haiti is part of this Administration's continuing efforts to support Haiti's recovery."[32]

FEDERAL ASSISTANCE TO HAITIAN MIGRANTS

Those Haitians who are deemed Cuban-Haitian Entrants are among the subset of foreign nationals who are eligible for federal benefits and cash assistance. Those Haitians who are newly arriving LPRs, however, are barred from the major federal benefits and cash assistance for the first five years after entry. Over a decade ago, Title IV of the Personal Responsibility and Work Opportunity Reconciliation Act (PRWORA) of 1996 (P.L. 104-193) established comprehensive restrictions on the eligibility of noncitizens for means-tested public assistance and limited the eligibility of refugees and asylees to five years.[33] Foreign nationals who enter the United States on temporary visas (i.e., nonimmigrants) and those who enter the United States without authorization are barred from any federal public benefit except the emergency services and programs expressly listed in PRWORA.[34] Amendments in P.L. 105-33 and P.L. 105-185 extended the period of food stamp/Supplemental Security Income (SSI)/Medicaid (but not Temporary Assistance for Needy Families) eligibility for refugees and asylees from five to seven years and added Cuban and Haitian Entrants to those eligible for these benefits for seven years.

Cuban-Haitian Entrants

The term "Cuban-Haitian Entrant" is not defined in the INA, but its usage dates back to 1980. Many of the Cubans and the vast majority of the Haitians who arrived in South Florida during the 1980 Mariel Boatlift did not qualify for asylum according to the individualized definition of persecution in §§ 207-208 of the INA. The Carter Administration labeled Cubans and Haitians as "Cuban-Haitian Entrants" and used the discretionary parole authority of the Attorney General to admit them to the United States. Subsequently, an adjustment of status provision was included in the Immigration Reform and Control Act (IRCA) of 1986 (P.L. 99-603) that enabled the Cuban-Haitian Entrants who had arrived during the Mariel Boatlift to become LPRs.[35]

While not a term of law in the INA, Congress did define Cuban-Haitian Entrant in the context of eligibility for federal assistance. Title V of the Refugee Education Assistance Act of 1980 (P.L. 96-422), commonly known as Fascell-Stone, defined Cuban and Haitian Entrants as:

> (1) any individual granted parole status as a Cuban/Haitian Entrant (Status Pending) or granted any other special status subsequently established under the immigration laws for nationals of Cuba or Haiti, regardless of the status of the individual at the time assistance or services are provided; and
>
> A. who—(i) was paroled into the United States and has not acquired any other status under the Immigration and Nationality Act [this chapter]; (ii) is the subject of removal proceedings under the Immigration and Nationality Act [this chapter]; or (iii) has an application for asylum pending with the Immigration and Naturalization Service; and
>
> B. with respect to whom a final, nonappealable, and legally enforceable order of removal has not been entered.[36]

The intent and resulting effect of the Fascell-Stone provision was to treat Cubans and Haitians in the same manner as refugees and asylees for the purposes of the federal refugee resettlement program and most other federal benefits and assistance.[37] The Office of Refugee Resettlement (ORR) uses the term Cuban-Haitian Entrants for Haitians who meet the Fascell-Stone definition. It could be argued that Haitians who currently benefit from TPS meet the definition of Cuban-Haitian Entrant, as do those who are paroled into the United States after the January 12, 2010, earthquake. It remains to be seen, however, whether policy makers will attempt to narrow the applicability of the Cuban-Haitian Entrants classification for Haitians displaced by the earthquake.[38]

Major Federal Benefit Programs

Supplemental Security Income (SSI)

Under current law, asylees, refugees, and Cuban-Haitian entrants (as well as certain aliens whose deportation/removal is being withheld for humanitarian reasons and Vietnam-born Amerasians fathered by U.S. citizens) are among categories of aliens who may be eligible for SSI *for seven years* after entry/grant of such status.[39] In order to receive SSI benefits, these qualified aliens must meet all the requirements for eligibility as native-born citizens. SSI eligibility requirements include meeting the definitions for age, disability, or blindness and falling below established income and resource thresholds.[40] This SSI eligibility for Cuban-Haitian entrants as well as refugees, asylees, and aliens in other specified humanitarian categories was extended to nine years (during FY2009 through FY2011) by P.L. 110-328. If qualified aliens are eligible for SSI, they are likely to be eligible for enrollment in their state's Medicaid Program.[41]

Medicaid

Haitians deemed to be Cuban-Haitian Entrants are eligible for Medicaid until they have been in the United States for seven years. After the initial seven years, states have the option to continue to provide Medicaid. After five years in the United States, Haitians who are LPRs may become eligible for Medicaid at the state's option. Those Haitian LPRs with a substantial work history— generally 10 years (40 quarters) of work documented by Social Security or other employment records—or a military connection (active duty military personnel, veterans, and their families) are also eligible.[42] As noted above, Medicaid coverage is required for all otherwise qualified SSI recipients, so long as they meet SSI noncitizen eligibility tests.

Temporary Assistance for Needy Families (TANF)

Cuban-Haitian Entrants are treated as refugees, and thus those families with children under 18 may be eligible for time-limited cash assistance through a state's Temporary Assistance for Needy Families (TANF) program. PRWORA restricted receipt of federal TANF benefits to a 60-month lifetime limit. States may exempt up to 20% of the caseload from the time limit due to state-defined hardship, and states have the option to continue TANF benefits under special circumstances. Like other federal welfare programs, the TANF program is means-tested; however, unlike SSI, the TANF program is state-administered and payment levels can vary widely by state.[43] Refugees and entrants participating in job training programs sponsored by the Office of Refugee Resettlement (ORR) are considered to be working toward self-sufficiency and may be exempt from certain state TANF program requirements. TANF beneficiaries may be eligible for their state's Medicaid program; however SSI beneficiaries are generally ineligible to receive TANF in addition to SSI.

Refugee Resettlement Assistance[44]

As noted above, those Haitians deemed to be Cuban-Haitian entrants are eligible for the federal resettlement assistance program for refugees and entrants, which is partially funded through the ORR. In addition to providing a range of social services, primarily administered by states, the ORR provides funding to states for transitional cash and medical assistance through the Transition and Medical Services program.[45] ORR resettlement assistance and services are designed to help refugees and entrants obtain self-sufficiency and social adjustment as quickly as possible. Refugees and entrants are expected to become self-sufficient within six months of arrival, and Refugee Cash Assistance (RCA) and Refugee Medical Assistance (RMA) are limited to eight months.

Refugees and entrants who meet the income and resource eligibility requirements for SSI, TANF or Medicaid, but are not otherwise eligible (e.g., single males or childless females and couples), may receive benefits under the ORR-funded RCA and RMA programs.[46] Under Title V of the Refugee Education Assistance Act, participating states are fully reimbursed for cash and medical assistance to Cuban and Haitian entrants under the same conditions and to the same extent as such assistance and services for refugees under the refugee program.

Refugee Cash Assistance (RCA)

States have the option of choosing either a publicly administered or public/private RCA program.[47] Most states publicly administer their RCA program. By doing so, the state agency must operate its RCA program consistent with the provisions of their TANF program.[48] In publicly administered RCA programs, payment levels to refugees or entrants are equivalent to a state's TANF payment levels, and thus vary across the country.

Refugee Medical Assistance (RMA)

Like RCA benefits, states administer their own RMA program from ORR funds which pay for 100% of RMA costs for eligible refugees and entrants. Refugees or entrants who are not eligible for their state's Medicaid program may be eligible for RMA benefits for up to eight months. In many states, covered services under RMA are the same as covered services under their state Medicaid plan. Costs of RMA per refugee vary due to variables such as age, health of the beneficiary, and services provided.

LEGISLATION IN 111TH CONGRESS

Haitian Family Members with Approved Petitions

Some U.S. citizens and legal permanent residents (LPRs) have family in Haiti for whom they have petitioned for visas to become LPRs in the United States.[49] According to the DOS, there are 54,716 Haitians who have approved petitions to immigrate to the United States and who are waiting for one of these numerically limited visas to become available.[50] The INA provides for a permanent annual worldwide level of 675,000 legal permanent residents (LPRs), but this level is flexible and certain categories of LPRs are permitted to exceed the limits. The INA establishes per-country levels at 7% of the worldwide level for other family-sponsored LPRs.[51] Immediate relatives of U.S. citizens are among those exempt from direct numerical limits.[52]

According to the INA, family-sponsored (and employment-based) preference visas are issued to eligible immigrants in the order in which a petition has been filed. Spouses and children of prospective LPRs are entitled to the same status, and the same order of consideration as the person qualifying as principal LPR, if accompanying or following to join (referred to as derivative status). When visa demand exceeds the per-country limit, visas are prorated according to the preference system allocations for the oversubscribed foreign state or dependent area. Relatives of U.S. citizens and LPRs are waiting in backlogs for a visa to become available, with the brothers and sisters of U.S. citizens now waiting about 11 years, with even longer waits for siblings from Mexico and the Philippines. Married adult sons and daughters of U.S. citizens who filed petitions almost nine years ago (May 22, 2001) are now being processed for visas. Haitians with family in the United States are in this worldwide backlog for visas.

Advocates for Haitians are asking Secretary Napolitano to give humanitarian parole to those Haitians with approved petitions for visas. In the context of immigration law, parole means that the foreign national has been granted temporary permission to be present in the United States. Parole does not constitute formal admission to the United States, and parolees are required to leave when the terms of their parole expire, or if otherwise eligible, to be admitted in a lawful status.[53] Options to expedite the immigration of Haitians with approved petitions that would require legislative actions would include enacting an amendment to the INA that exempts certain Haitians for the numerical limits, and amending a transitional nonimmigrant visa—the V visa— for immediate relatives (spouse and children) of LPRs who have had petitions to also become LPRs pending for three years to include certain Haitians.[54]

Proponents of expediting the admission of Haitians with family in the United States maintain that it would relieve at least some of the humanitarian burden in Haiti. "Unless our government does something to make sure loved ones can join families quickly," argues Cheryl Little of the Florida Immigrant Advocacy Center, "it could have devastating consequences for those in Haiti waiting for these visas."[55] Other supporters assert that it would increase the remittances sent back to Haiti to provide critical help as the nation tries to rebuild.[56] "A larger Haitian diaspora would be a far better base for the country's economic future than aid pledges that may or may not be met," according to Elliot Abrams of the Council on Foreign Relations.[57] Those opposed to expediting the admission of Haitians assert that it would not be in the national interest, nor would it be fair to other foreign nationals waiting to reunite with their families. The Center for Immigration Studies' Mark Krikorian has gone on record saying that "poverty and underdevelopment can't be criteria we use to pick immigrants," and concludes that the "place to help Haitians is in Haiti, not the United States."[58]

The Haitian Emergency Life Protection Act of 2010 (S. 2998/H.R. 4616) would amend the INA to allow Haitian nationals whose petition for a family-sponsored immigrant visa was approved on or before January 12, 2010, to obtain nonimmigrant visas under §101(a)(15)(V). As noted earlier, this nonimmigrant visa—known as the V visa—is a transitional visa category for immediate relatives (spouse and children) of LPRs who have had petitions to also become LPRs pending for three years.

Adoption of Haitian Orphans

Haitian children who were legally confirmed as orphans eligible for intercountry adoption by the government of Haiti and who were in the process of being adopted by U.S. residents prior to the earthquake have been given humanitarian parole to come to the United States. DHS Secretary Napolitano's announcement of the policy on January 18, 2010, indicated that the Haitian orphans must meet one of following criteria to be eligible:

> Evidence of availability for adoption, which includes at least a full and final Haitian adoption decree, a Government of Haiti custody grant to prospective adoptive parents for emigration and adoption, or secondary evidence in lieu of the first two criteria.
>
> Evidence of suitability for adoption, which includes at least a Notice of Approval of Form I-600A, *Application for Advance Processing of an Orphan Petition,* a current FBI

fingerprints and background security check clearances, or physical custody in Haiti plus a security background check.

Other Haitian orphans potentially eligible for humanitarian parole include children who were identified by an adoption service provider or facilitator as eligible for intercountry adoption and who were matched to prospective American adoptive parents prior to January 12, 2010. When it announced the humanitarian parole for Haitian orphans, DHS acknowledged, "Given the severity of the disaster in Haiti, we understand that there are additional children that have been orphaned and/or separated from relatives and may also be in varying stages of the adoption process. DHS and the U.S. Department of State continue to evaluate additional eligibility criteria and will provide additional information as soon as it is available."[59] No policy or procedures, however, have been announced regarding prospective adoptions of Haitian children orphaned as a result of the earthquake. [60] As of March 29, 2010, the United States has given humanitarian parole to 1,050 Haitian orphans, and 902 of those have already arrived in the United States.[61] The program to grant humanitarian parole to enter the United States to certain Haitian children who were in the process of being adopted by U.S. residents prior to the earthquake ended in April 2010. Under most circumstances, when adoptive children immigrate to the United States, they are admitted as LPRs.[62] The children granted parole under this program would otherwise have to live with their families in the United States for two years before they will be eligible to become LPRs.

On July 20, 2010, the House passed the Help HAITI Act of 2010 (H.R. 5283), which would authorize the DHS Secretary to adjust to LPR status those aliens who were granted parole into the United States pursuant to the humanitarian parole policy for certain Haitian orphans from January 18, 2010, through April 15, 2010. A similar bill, S. 3411, was introduced in the Senate. To qualify for the adjustment, the applicant must be physically present in the United States when the adjustment application is filed and be admissible as an LPR. The requirements are applicable to adopted children if, before the alien is 18 years of age, he or she adjusts LPR status and is adopted by a U.S. citizen (which may occur before, on, or after status adjustment). The legislation would permit a parent or legal guardian to apply on behalf of a minor and would prohibit any derivative immigration benefits for the birth parent of an alien adjusted under this act. The Senate passed H.R. 5283 with an amendment on August 4, 2010. On December 1, 2010, the House suspended the rules and agreed to the Senate amended version of H.R. 5283, which became P.L. 111-293 on December 9, 2010. P.L. 111-293 made the children who were paroled into the country as Haitian orphans eligible to immediately adjust to LPR status, and other provisions of immigration law enable them to automatically become U.S. citizens when granted LPR status.

Haitian Orphan Placement Effort Act of 2010 (H.R. 4603) would have directed the DHS Secretary to expand the humanitarian parole policy for certain Haitian orphans announced on January 18, 2010, so it would have applied on a case-by-case basis to children who were legally confirmed as orphans eligible for intercountry adoption by the government of Haiti before January 12, 2010. The bill also would have authorized the placement of Haitian children granted humanitarian parole into the United States in an unaccompanied refugee minor program if a suitable family member is not available to provide care.

Supplemental FY2010 Funding

President Barack Obama requested that the Congress consider supplemental FY2010 funding to provide for costs associated with relief and reconstruction support for Haiti, including reimbursement of obligations that have already been incurred by federal agencies after the January 12, 2010, earthquake.[63] Only a small portion of the $2.8 billion requested would pertain to Haitian evacuees and migrants in the United States. Specifically, the President requested $220.0 million for the Department of Health and Human Services (HHS) to fund four types of activities, of which two are directly related to Haitians brought to the United States after the earthquake. [64] These two are the state share of Medicaid and Children's Health Insurance Program (CHIP) costs for eligible Haitians; and cash, medical, and repatriation assistance for eligible Haitians.[65] The request does not specify how much funding would be allocated to each of these activities. The President's request did not propose any changes or expansions in eligibility for assistance or benefits as described above.[66] The Supplemental Appropriations Act, 2010 (H.R. 4899, P.L. 111-212), provides the requested $220 million.

The President's supplemental request also included $15 million for the USCIS Examinations Fee Account.[67] USCIS funds the processing and adjudication of immigrant, nonimmigrant, refugee, asylum, and citizenship benefits almost entirely through monies generated by the Examinations Fee Account.[68] USCIS charges fees for almost all adjudications and services; however, the agency traditionally has not charged the Examination Fee for refugees and asylum seekers. The Administration proposed to use funds for reception and settlement services provided to designated Haitians; fee waivers for eligible Haitians granted TPS; humanitarian parole to bring medical evacuees and certain categories of Haitians into the United States; and costs associated with adoptions and orphans. The Supplemental Appropriations Act, 2010 (H.R. 4899, P.L. 111- 212), provides $10.6 million to the USCIS for costs associated with processing the Haitian migrants.[69]

Status Adjustment

Several versions of the legislation on comprehensive immigration reform that stalled in the Senate in June 2007 (e.g., S. 1348 and S. 1639) included provisions that would have enabled many of the Haitians in the United States without authorization to adjust to LPR status under certain circumstances and with some penalties. In the 110th Congress, H.R. 1645 also included provisions that would have allowed HRIFA adjustments to encompass a child of an applicant based on the child's age and status on October 21, 1998. H.R. 750 would have, among other things, authorized the adjustment of status for certain nationals or citizens of Haiti who are present in the United States. H.R. 454 would have amended HRIFA to provide that determinations with respect to children be made according to their age and status as of October 21, 1998; would have permitted an application based upon child status to be filed by a parent or guardian if the child is present in the United States on such filing date; and would have included document fraud among the grounds of inadmissibility, which would not have precluded an otherwise qualifying Haitian alien from permanent resident status adjustment. Many of these elements are included in a comprehensive immigration reform piece of legislation in the 111th Congress (H.R. 4321).

During the 110th Congress, §105 of the FY2008 Consolidated Appropriations Act (P.L. 110-161) continued the prohibition of the use of funds to provide visas to certain aliens who were involved in political violence in Haiti.

Haitian Migration Issues in the 112th Congress

Haitians with Approved Visa Petitions and the Impact of the Cholera Outbreak

As noted above, U.S. citizens and LPRs have family in Haiti for whom they have petitioned for visas to become LPRs in the United States. Last year, the Government of Canada established the Haiti Special Measures Program to expedite the processing of Haitians with family members in Canada and reported processing more than five times the number of applications in 2010 that were processed in 2009 for the same time frame.[70] Some have called on DHS Secretary Napolitano to use the humanitarian parole authority under the INA to allow Haitians with approved visa petitions to enter the United States without waiting for visas to become available. Those opposed to expediting the admission of Haitians argue that it would not be fair to other foreign nationals waiting to reunite with their families.

The cholera outbreak, which CDC confirmed on October 21, 2010, has renewed appeals to Congress to enact legislation that would expedite the admission of Haitians with family in the United States. Although some concerns are being raised about the possible spread of cholera by Haitians coming to the United States, others point out that medical examinations and health screenings are core features of U.S. immigration policy.[71] Supporters of the current policy assert that expediting the admission of Haitians would not be in the U.S. national interest at this time, especially given the cholera outbreak.

Possible Mass Migration

There are concerns that the ongoing crisis in Haiti may result in mass migration from the country. Not only has there been massive displacement of people caused by the earthquake, but observers of the situation warn of potential and widespread lawlessness as well as outbreaks of disease. These health, safety, and security factors—individually or in combination—could trigger an exodus of Haitians seeking refuge in nearby countries, including the United States. Moreover, the unexpected return to Haiti on January 16, 2011, of Jean-Claude "Baby Doc" Duvalier, the dictator deposed 25 years ago, has heightened fears that political turmoil may ensue. The practice of Haitians fleeing by the thousands began when Duvalier was dictator.[72]

At least five federal agencies now handle Haitian migrants: DHS's Coast Guard (interdiction); Customs and Border Protection (apprehensions and inspections); Immigration and Customs Enforcement (detention); U.S. Citizenship and Immigration Services (credible fear determination); and DOJ's EOIR (asylum and removal hearings). DHS would take the lead in handling a potential mass migration and has long had a set of operational plans in place to respond to such a situation. In her TPS announcement, Secretary Napolitano warned of the consequences of Haitians fleeing to the United States:

> At this moment of tragedy in Haiti it is tempting for people suffering in the aftermath of the earthquake to seek refuge elsewhere. But attempting to leave Haiti now will only bring more hardship to the Haitian people and nation.... It is important to note that TPS will apply only to those individuals who were in the United States as of January 12, 2010. Those who attempt to travel to the United States after January 12, 2010 will not be eligible for TPS and will be repatriated. [73]

The balancing of DHS's border security and immigration control responsibilities during an ongoing humanitarian crisis poses a unique challenge.

Resumption of Deportations

The DHS Immigration and Customs Enforcement (ICE) bureau deported 27 Haitians, reportedly 26 of whom had criminal convictions and one of whom was deemed a national security risk.[74] ICE had announced in December 2010 that it was resuming the Haitian deportations that were halted immediately following the 2010 earthquake. Reportedly 300-350 Haitians with criminal convictions are currently in custody and slated for removal. In response to questions about who among those Haitians with orders of removal were actually scheduled for deportation, ICE spokeswoman Barbara Gonzalez stated that the decisions would be made "consistent with our domestic immigration enforcement priorities."[75]

Advocates for the Haitians point out that an estimated 1.3 million Haitians remain displaced from their homes after the 2010 earthquake and that the forced repatriation of Haitians would fuel the political turmoil in Haiti. Some argue further that deportation would endanger the lives of the deportees because the Haitian government often detains criminals who have been repatriated, and the Haitian jails are riddled with cholera. ICE, however, maintains that deportation is preferable to releasing large numbers of people with criminal convictions back into U.S. communities. An ICE official said the State Department had been working with Haitian officials "to ensure that the resumption of removals is conducted in a safe, humane manner with minimal disruption to ongoing rebuilding efforts."[76]

End Notes

[1] For further discussion on current conditions in Haiti, see CRS Report R40507, *Haiti: Current Conditions and Congressional Concerns*, by Maureen Taft-Morales, and CRS Report R41023, *Haiti Earthquake: Crisis and Response*, by Rhoda Margesson and Maureen Taft-Morales.

[2] Centers for Disease Control and Prevention, *Cholera in Haiti*, U.S. Department of Health and Human Service, Travel Health Precaution, January 12, 2011, http://wwwnc.cdc.gov/travel/content/travel-health

[3] During a seven-month period in 1980, approximately 125,000 Cubans and 25,000 Haitians arrived by boats to South Florida. This mass migration became known as the Mariel boatlift because most of the Cubans departed from Mariel Harbor in Cuba.

[4] For historical analysis of conditions in Haiti, see CRS Report RL32294, *Haiti: Developments and U.S. Policy Since 1991 and Current Congressional Concerns*, by Maureen Taft-Morales and Clare Ribando Seelke.

[5] For interdiction data, see http://www.uscg.mil/hq/cg5/cg531/AMIO/FlowStats/currentstats.asp

[6] Manuel Roig-Franzia, "20 Haitian Migrants Die at Sea; 58 Missing," *The Washington Post*, May 5, 2007, A12.

[7] CRS congressional distribution memorandum, *Estimating the U.S. Haitian Population by State*, by William Kandel, January 22, 2010. (available by request)

[8] Using the 2008 ACS rather than the 2006-2008 pooled ACS, other researchers report an estimate of 535,000 Haitian-born residents. Aaron Terrazas, "Haitian Immigrants in the United States," *Migration Information Source*, January, 2010.

[9] Department of Homeland Security Office of Immigration Statistics, *2008 Yearbook of Immigration Statistics*, August 2009.

[10] Department of Homeland Security, Office of Immigration Statistics, *Estimates of the Legal Permanent Resident Population in 2008*, by Nancy Rytina, 2009.

[11] Department of Homeland Security, Office of Immigration Statistics, *Estimates of the Unauthorized Immigrant Population Residing in the United States: January 2008*, by Michael Hoefer, Nancy Rytina, and Bryan C. Baker, 2009.

[12] Telephone conversation with USCIS Office of Legislative Affairs, March 30, 2010.

[13] Aliens must demonstrate a well-founded fear that if returned home, they will be persecuted based upon one of five characteristics: race, religion, nationality, membership in a particular social group, or political opinion.

[14] §202 of the Immigration Reform and Control Act of 1986 (P.L. 99-603).

[15] CRS Report 93-233, *Asylum Seekers: Haitians in Comparative Context*, by Ruth Ellen Wasem. (Archived report available upon request.)

[16] *Miami Herald*, "U.S. Policy Unjust to Haitians Fleeing Violence," January 9, 2006; data confirmed in telephone conversation with DHS officials, January 12, 2006.

[17] CRS Report 98-270, *Immigration: Haitian Relief Issues and Legislation*, by Ruth Ellen Wasem. (Archived report available upon request.)

[18] CRS Report RL33109, *Immigration Policy on Expedited Removal of Aliens*, by Alison Siskin and Ruth Ellen Wasem.

[19] U.S. Department of Justice, Executive Office for Immigration Review, *FY2006 Statistical Yearbook.*

[20] U.S. Department of Justice, Executive Office for Immigration Review, *FY2008 Statistical Yearbook.*

[21] Communication between DHS Immigration and Customs Enforcement and the Congressional Research Service, January 19, 2010.

[22] *Federal Register*, vol. 67, no. 219, pp. 68923-68926 (November 13, 2002).

[23] 23 I&N December 572 (A.G. 2003).

[24] CRS Report RL32369, *Immigration-Related Detention: Current Legislative Issues*, by Chad C. Haddal and Alison Siskin.

[25] Letter from Daniel J. Bryant, Assistant Attorney General, to Sens. Edward Kennedy and Sam Brownback, dated September 25, 2002.

[26] U.S. Senate, Committee on the Judiciary, Subcommittee on Immigration, *Hearing on the Detention and Treatment of Haitian Asylum Seekers*, October 1, 2002. (Hereafter cited as Senate Subcommittee on Immigration, *Hearing on Haitian Asylum Seekers*.)

[27] George Lardner Jr., "More Illegal Immigrants Can Be Held Ashcroft's Ruling Cites National Security Issues," *The Washington Post*, April 25, 2003, p. A6.

[28] Senate Subcommittee on Immigration, *Hearing on Haitian Asylum Seekers*

[29] For additional information on Temporary Protected Status, see CRS Report RS20844, *Temporary Protected Status: Current Immigration Policy and Issues*, by Ruth Ellen Wasem and Karma Ester.

[30] Trenton Daniel and Jacqueline Charles, "Mud and Misery Rule Storm-Ravaged City," *Miami Herald*, November 2, 2008, p. A1.

[31] Toluse Olorunnipa and Alfonso Chardy, "U.S. Halts Deportation of Undocumented Haitians Due to Earthquake," *Miami Herald*, January 13, 2010.

[32] U.S. Department of Homeland Security, "Statement from Secretary Janet Napolitano," press release, January 15, 2010.

[33] CRS Report RL33809, *Noncitizen Eligibility for Federal Public Assistance: Policy Overview and Trends*, by Ruth Ellen Wasem.

[34] CRS Report RL34500, *Unauthorized Aliens' Access to Federal Benefits: Policy and Issues*, by Ruth Ellen Wasem.

[35] § 202 of the Immigration Reform and Control Act of 1986 (P.L. 99-603).

[36] §501(e)(1) of the Refugee Education Assistance Act of 1980 (P.L. 96-422).

[37] §501(a)(1) of P.L. 96-422 states: "The President shall exercise authorities with respect to Cuban and Haitian entrants which are identical to the authorities which are exercised under chapter 2 of title IV of the Immigration and Nationality Act."

[38] For a discussion of how the classification of Cuban-Haitian Entrant is applied to Cubans in the United States, see CRS Report R40566, *Cuban Migration to the United States: Policy and Trends*, by Ruth Ellen Wasem.

[39] In most cases, an officer of the U.S. Citizenship and Immigration Services from DHS determines if a non-citizen is granted status in one of the qualified alien categories.

[40] Countable resource limit for SSI eligibility is $2,000 for individuals and $3,000 for couples. See CRS Report RS20294, *Supplemental Security Income (SSI): Beneficiary Income/Resource Limits and Accounts Exempt from Benefit Determinations*, by Umar Moulta-Ali for additional information on income and resource limits and exclusions.

[41] Thirty-two states link federal eligibility for SSI with state-administered Medicaid eligibility by automatically qualifying SSI recipients for Medicaid. Other states either follow the Social Security Administration's (SSA) SSI eligibility rules and require a separate Medicaid application, or they establish their own Medicaid eligibility rules outside of the SSA. For more information, see the Social Security Administration's Medicaid Information at http://www.socialsecurity.gov/disabilityresearch/wi/medicaid.htm.

[42] CRS Report R40144, *State Medicaid and CHIP Coverage of Noncitizens*, by Ruth Ellen Wasem.

[43] In some states, TANF applications are filed at the county level and benefit levels can vary by county.

[44] For a fuller discussion of refugee resettlement, see CRS Report RL31269, *Refugee Admissions and Resettlement Policy*, by Andorra Bruno.

[45] ORR cannot reimburse states for TANF, SSI, or Medicaid programs.

[46] U.S. Department of Health and Human Services, Administration for Children and Families, *Annual ORR Reports to Congress-2005*, at http://www.acf.hhs.gov/programs/orr/data/arc.htm, accessed on February 24, 2008.

[47] In 2000, ORR published a final rule amending the requirements governing RCA. States were given a choice in how RCA services would be administered in their states. Codified at 45 CFR Part 400.

[48] *Annual ORR Reports to Congress-2005.*

[49] Petitions for LPR status are first filed with USCIS by the sponsoring relative or employer in the United States. If the prospective immigrant is already residing in the United States, the USCIS handles the entire process, which is called "adjustment of status" because the alien is moving from a temporary category to LPR status. If the prospective LPR does not have legal residence in the United States, the petition is forwarded to the DOS Bureau of Consular Affairs in their home country after USCIS has reviewed it. CRS Report RL32235, *U.S. Immigration Policy on Permanent Admissions*, by Ruth Ellen Wasem.

[50] U.S. Department of State, *Annual Report of Immigrant Visa Applicants in the Family-sponsored and Employment-based preferences Registered at the National Visa Center as of November 1, 2009*, http://www.travel.stateWaitingListItem.pdf.

[51] The numerically limited family preference categories are (1) unmarried sons and daughters of citizens, (2) spouses and children of LPRs, (3) unmarried sons and daughters of LPRs, (4) married sons and daughters of citizens, and (5) siblings of citizens age 21 and over.

[52] "Immediate relatives" are defined by the INA to include the spouses and unmarried minor children of U.S. citizens, and the parents of adult U.S. citizens.

[53] §212(d)(5)(A) of the INA.

[54] For further information on the V visas and other nonimmigrant visa categories, see CRS Report RL31381, *U.S. Immigration Policy on Temporary Admissions*, by Chad C. Haddal and Ruth Ellen Wasem.

[55] William Gibson, "Clinton: More immigrants from Haiti may be allowed," *Sun Sentinel*, January 28, 2010.

[56] Michael Clemons, "To help Haiti's earthquake victims, change U.S. immigration laws," *Washington Post*, January 24, 2010.

[57] Elliot Abrams, "What Haiti Needs—A Haitian Diaspora," *Washington Post*, January 22, 2010.

[58] Amy Goldstein and Peter Whoriskey, "Debate grows in aftermath of quake: Should U.S. let more Haitians immigrate?," *Washington Post*, January 25, 2010; and Mark Krikorian , "Help Haitians—in Haiti," *National Review Online*, January 22, 2010.

[59] U.S. Department of Homeland Security, " Secretary Napolitano Announces Humanitarian Parole Policy for Certain Haitian Orphans," fact sheet, January 18, 2010.

[60] International adoption is a two-step process. First, the parents' eligibility to adopt must be verified, and then once the child is identified and the parents have complied with the laws of the sending country, the adoptive parents apply for a visa for the child so that the child can legally immigrate to the United States. The application for the visa triggers an investigation into the child's background to confirm that the child has not been bought or

stolen, and meets the definition of orphan in the INA. USCIS verifies the eligibility to adopt while the DOS processes the visa application for the child. Once the prospective parents have been deemed eligible to adopt, USCIS policy states they have 18 months to complete the adoption. CRS Report RL31769, *Immigration: International Child Adoption*, by Alison Siskin

[61] Telephone conversation with USCIS Office of Legislative Affairs, March 30, 2010.

[62] Under the Child Citizenship Act of 2000 (CCA; P.L. 106-395), most children adopted by U.S. citizens become U.S. citizens upon admission to the United States.

[63] Letter from Barack H. Obama, President of the United States, to Nancy Pelosi, Speaker of the United States House of Representatives, March 24, 2010, http://www.whitehouse.gov/omb/assets amendment_03_24_10.pdf.

[64] The other two activities included in $220 million request for HHS would cover costs associated with medical evacuations and costs for HHS public health activities in Haiti.

[65] In a telephone conversation March 30, 2010, the USCIS Office of Legislative Affairs reported that 215 Haitians had been granted humanitarian parole as of March 29, 2010 (in addition to the Haitian orphans who had received parole).

[66] Letter from Peter R. Orszag, Office of Management and Budget Director, to Barack Obama, President of the United States, March 24, 2010, http://www.whitehouse.gov/omb/assets/ budget_amendments/ amendment_03_24_10.pdf.

[67] Letter from Peter R. Orszag, Office of Management and Budget Director, to Barack Obama, President of the United States, March 24, 2010, http://www.whitehouse.gov/omb/assets/budget_amendments/ amendment_03_24_10.pdf.

[68] CRS Report RL34040, *U.S. Citizenship and Immigration Services' Immigration Fees and Adjudication Costs: Proposed Adjustments and Historical Context*, by William A. Kandel.

[69] For a full accounting of the FY2010 Supplemental, see CRS Report R41232, *FY2010 Supplemental for Wars, Disaster Assistance, Haiti Relief, and Other Programs*, coordinated by Amy Belasco.

[70] Citizenship and Immigration Canada, *Citizenship and Immigration Canada Secures Fast Family Reunification for Haitians Affected by Earthquake*, News Release, July 13, 2010, http://www.cic.gc.ca/english/department/ media/releases/2010/2010-07-13b.asp.

[71] For a full discussion, see CRS Report R40570, *Immigration Policies and Issues on Health-Related Grounds for Exclusion*, by Ruth Ellen Wasem.

[72] When he became dictator in 1971, Jean-Claude Duvalier continued the legacy of repression and corruption begun in 1957 when his father Francois "Papa Doc" Duvalier took over Haiti. Under their rule, arbitrary imprisonment, torture, and unexplained deaths became commonplace. The Duvaliers' private militia, the Tontons Macoutes, carried out most of this repression and also served to counterbalance the army's power, which the Duvaliers kept in check to prevent military coups. For further background, see CRS Report 93-931, *Haiti: Background to the 1991 Overthrow of President Aristide*, by Maureen Taft-Morales.

[73] U.S. Department of Homeland Security, "Statement from Secretary Janet Napolitano," press release, January 15, 2010.

[74] "US Deports First Haitians since Earthquake," *The Associate Press*, January 21, 2011.

[75] Kirk Semple, "Haitians in U.S. Brace for Deportations to Resume," *The New York Times*, December 19, 2010; Greg Allen, "U.S. To Deport Haitians With Criminal Records," *National Public Radio Morning Edition*, January 12, 2011.

[76] Greg Allen, "U.S. To Deport Haitians With Criminal Records," *National Public Radio Morning Edition*, January 12, 2011; "Fighting Haitian Deportations from Florida ," *Associated Press*, January 6, 2011; and, Kirk Semple, "Haitians in U.S. Brace for Deportations to Resume," *The New York Times*, December 19, 2010.

In: Haiti: Earthquake and Response
Editor: Rachael A. Donlon
ISBN: 978-1-61668-998-8

Chapter 4

EARTHQUAKES: RISK, DETECTION, WARNING, AND RESEARCH*

Peter Folger

SUMMARY

The 1994 Northridge (CA) earthquake caused as much as $26 billion (in 2005 dollars) in damage and was one of the costliest natural disasters to strike the United States. The Federal Emergency Management Agency has estimated that earthquakes cost the United States over $5 billion per year. A hypothetical scenario for a magnitude 7.8 earthquake in southern California estimated a possibility of 1,800 fatalities and over $200 billion in economic losses. The January 12, 2010, magnitude 7.0 earthquake that struck Haiti only 15 miles from Port-au-Prince, the capital city, has caused an estimated 230,000 fatalities and 300,000 injuries. On February 27, 2010, a magnitude 8.8 earthquake occurred offshore of Chile and has caused massive damage, although far fewer fatalities than the Haiti earthquake. The Cascadia Subduction Zone megathrust fault, which stretches from southern British Columbia to northern California, has the potential to generate a massive earthquake and tsunami, similar to the Chilean earthquake, that could threaten Oregon, Washington, and northern California.

The United States faces the possibility of large economic losses from earthquake-damaged buildings and infrastructure. California alone accounts for most of the estimated annualized earthquake losses for the nation, and with Oregon and Washington the three states account for nearly $4.1 billion (77%) of the U.S. total estimated annualized loss. A single large earthquake, however, can cause far more damage than the average annual estimate.

An ongoing issue for Congress is whether the federally supported programs aimed at reducing U.S. vulnerability to earthquakes are an adequate response to the earthquake hazard. Under the National Earthquake Hazards Reduction Program (NEHRP), four federal agencies have responsibility for long-term earthquake risk reduction: the U.S. Geological Survey (USGS), the National Science Foundation (NSF), the Federal Emergency Management Agency (FEMA), and the National Institute of Standards and Technology (NIST). They variously assess U.S. earthquake hazards, send notifications of

---

* This is an edited, reformatted and augmented version of a Congressional Research Services publication, dated April 19, 2010.

seismic events, develop measures to reduce earthquake hazards, and conduct research to help reduce overall U.S. vulnerability to earthquakes.

Congress reauthorized NEHRP in 2004 (P.L. 108-360) through FY2009. Appropriations for NEHRP from FY2005 to FY2009 did not meet authorized levels; the total funding enacted was $613.2 million, approximately 68% of the total amount of $902.4 million authorized by P.L. 108- 360. The American Recovery and Reinvestment Act (ARRA; P.L. 111-5) provided some additional funding for earthquake activities under NEHRP through FY2010. In the 111th Congress, H.R. 3820 would reauthorize NEHRP through FY2014, authorizing total appropriations of $906 million over five years for the program, with 90% of the funding designated for the USGS and NSF, and the remainder for FEMA and NIST. The total authorized amounts are slightly greater than what was authorized by P.L. 108-360 over five years. If future appropriations match the funding levels authorized under H.R. 3820, these funds would exceed the total cumulative amounts actually appropriated between FY2005 and FY2009.

What effect funding at the levels enacted through FY2009 under NEHRP has had on the U.S. capability to detect earthquakes and minimize losses after an earthquake occurs is not clear. It is also difficult to predict precisely how NEHRP reauthorized under H.R. 3820 would achieve a major goal of the bill: to reduce the loss of life and damage to communities and infrastructure through increasing the adoption of hazard mitigation measures. A perennial issue for Congress is whether activities under NEHRP can reduce the potential for catastrophic loss in the next giant earthquake to strike the United States.

INTRODUCTION

Close to 75 million people in 39 states face some risk from earthquakes. Earthquake hazards are greatest in the western United States, particularly California, but also Alaska, Washington, Oregon, and Hawaii. Earthquake hazards are also prominent in the Rocky Mountain region and the New Madrid Seismic Zone (a portion of the central United States), as well as portions of the eastern seaboard, particularly South Carolina. Given the potentially huge costs associated with a large, damaging earthquake in the United States, an ongoing issue for Congress is whether the federally supported earthquake programs are appropriate for the earthquake risk.

Under the National Earthquake Hazards Reduction Program (NEHRP), the federal government supports efforts to assess and monitor earthquake hazards and risk in the United States. Four federal agencies, responsible for long-term earthquake risk reduction, coordinate their activities under NEHRP: the U.S. Geological Survey (USGS), the National Science Foundation (NSF), the Federal Emergency Management Agency (FEMA), and the National Institute of Standards and Technology (NIST). Congress reauthorized NEHRP in 2004 (P.L. 108-360), and authorized appropriations through FY2009 for a total of $902.4 million over five years. In the 111th Congress, H.R. 3820 (Title I) would reauthorize NEHRP through FY2014.

This report discusses:

- NEHRP—the multi-agency federal program to reduce the nation's risk from earthquakes;
- earthquake hazards and risk in the United States;
- federal programs that support earthquake monitoring;

- the U.S. capability to detect earthquakes and issue notifications and warnings; and
- federally supported research to improve the fundamental scientific understanding of earthquakes with a goal of reducing U.S. vulnerability.

National Earthquake Hazards Reduction Program (NEHRP)

In 1977 Congress passed the Earthquake Hazards Reduction Act (P.L. 95-124) establishing NEHRP as a long-term earthquake risk reduction program for the United States. The program initially focused on research, led by USGS and NSF, toward understanding and ultimately predicting earthquakes. Earthquake prediction has proved intractable thus far, and the NEHRP program shifted its focus to minimizing losses from earthquakes after they occur. FEMA was created in 1979 and President Carter designated it as the lead agency for NEHRP. In 1980, Congress reauthorized the Earthquake Hazards Reduction Act (P.L. 96-472), defining FEMA as the lead agency and authorizing additional funding for earthquake hazard preparedness and mitigation to FEMA and the National Bureau of Standards (now NIST).

A Shift in Program Emphasis to Hazard Reduction

In 1990, Congress reauthorized NEHRP (P.L. 101-614) and made substantive changes, to decrease the emphasis on earthquake prediction, clarify the role of FEMA, clarify and expand the program objectives, and require federal agencies to adopt seismic safety standards for new and existing federal buildings. In 2004, Congress reauthorized NEHRP through FY2009 (P.L. 108- 360) and shifted primary responsibility for planning and coordinating NEHRP from FEMA to NIST. It also established a new interagency coordinating committee and a new advisory committee, both focused on earthquake hazards reduction.

The current program activities are focused on four broad areas:

- developing effective measures to reduce earthquake hazards;
- promoting the adoption of earthquake hazards reduction measures by federal, state, and local governments, national building standards and model building code organizations, engineers, architects, building owners, and others who play a role in planning and constructing buildings, bridges, structures, and critical infrastructure or lifelines;
- improving the basic understanding of earthquakes and their effects on people and infrastructure, through interdisciplinary research involving engineering, natural sciences, and social, economic, and decision sciences; and
- developing and maintaining the Advanced National Seismic System (ANSS), the George E. Brown Jr. Network for Earthquake Engineering Simulation (NEES), and the Global Seismic Network (GSN).[1]

The House Science Committee report in the 108th Congress on H.R. 2608 (P.L. 108-360) noted that NEHRP has produced a wealth of useful information since 1977, but it also stated that the program's potential has been limited by the inability of the NEHRP agencies to coordinate their efforts.[2] The committee asserted that restructuring the program with NIST as the lead agency, directing funding towards appropriate priorities, and implementing it as a true interagency program would lead to improvement.

The 2004 reauthorization directed the Director of NIST to chair the Interagency Coordinating Committee. Other members of the committee include the directors of FEMA, USGS, NSF, the Office of Science and Technology Policy, and the Office of Management and Budget. The Interagency Coordinating Committee is charged with overseeing the planning, management, and coordination of the program. Primary responsibilities for the NEHRP agencies break down as follows (see also **Figure 1**):

- NIST supports the development of performance-based seismic engineering tools and works with other groups to promote the commercial application of the tools through building codes, standards, and construction practices.
- FEMA assists other agencies and private-sector groups to prepare and disseminate building codes and practices for structures and "lifelines",[3] and aids development of performance-based codes for buildings and other structures.
- USGS conducts research and other activities to characterize and assess earthquake risks, and (1) operates a forum, using the National Earthquake Information Center (NEIC), for the international exchange of earthquake information, (2) works with other NEHRP agencies to coordinate activities with earthquake reduction efforts in other countries, and (3) maintains seismic hazard maps in support of building codes for structures and lifelines, and other maps needed for performance-based design approaches.
- NSF supports research to improve safety and performance of buildings, structures, and lifelines using the large-scale experimental and computational facilities of NEES and other institutions engaged in research and implementation of NEHRP.

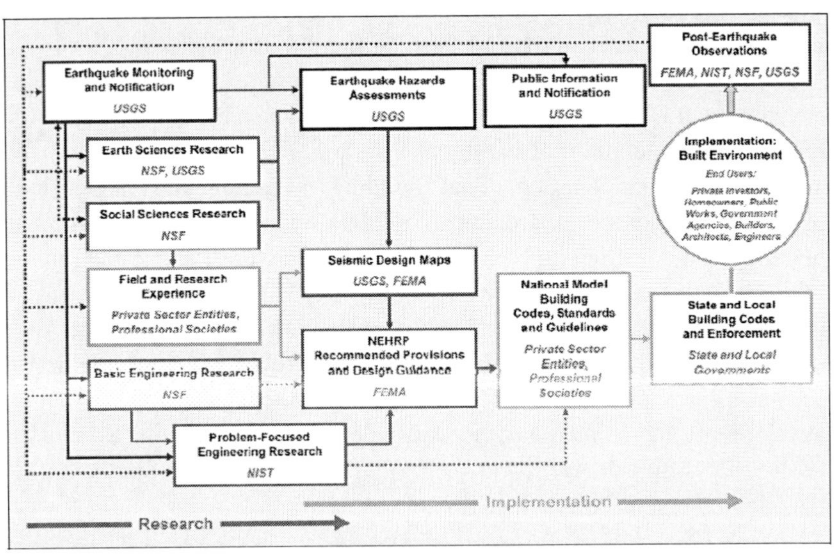

Source: NEHRP program office at http://www.nehrp.gov/pdf/ppt_sdr.pdf (modified by CRS).

Figure 1. NEHRP Agency Responsibilities and End Users of NEHRP Outcomes.

Table 1 shows the authorized and enacted appropriations for NEHRP from FY2005 through FY2010. The total enacted amount for FY2005-FY2009 was $613.2 million, or 68% of the $902.4 million total amount authorized in P.L. 108-360 over the five-year span. President Obama requested a total of $129.3 million for NEHRP in FY2011, even though authorization for the program under P.L. 108-360 expired at the end of FY2009.

Table 1. Authorized and Enacted Funding for NEHRP ($ millions)

		USGS	NSF	FEMA	NIST	Total
FY2005	Authorized	77.0	58.0	21.0	10.0	166.0
	Enacted	58.4	53.1	14.7	0.9	127.1
FY2006	Authorized	84.4	59.5	21.6	11.0	176.5
	Enacted	54.5	53.8	9.5	0.9	118.7
FY2007	Authorized	85.9	61.2	22.3	12.1	181.5
	Enacted	55.4	54.2	7.2	1.7	118.5
FY2008	Authorized	87.4	62.9	23.0	13.3	186.6
	Enacted	58.1	53.6	6.1	1.7	119.5
FY2009	Authorized	88.9	64.7	23.6	14.6	191.8
	Enacted	61.2	55.0	9.1	4.1	129.4
FY2010	Enacted	62.8	55.3	8.9	4.1	131.1
FY2011	Requested	62.3	53.8	9.1	4.1	129.3

Source: NEHRP program office, at http://www.nehrp.gov/pdf/ppt_budget_fy11.pdf.

Notes: According to the NEHRP program office, ARRA funds are not included; and the FY2011 requested budget is the estimated portion of the President's budget request that would be allocated for NEHRP activities. Also, the FY2010 enacted amounts are estimates.

NEHRP Reauthorization in the 111th Congress

Title I of H.R. 3820, the Natural Hazards Risk Reduction Act of 2009, would reauthorize NEHRP through FY2014, retain NIST as the lead NEHRP agency, and authorize total appropriations of about $906 million over five years. Title II of H.R. 3820 would reauthorize the National Windstorm Impact Reduction Act (first enacted in 2004 as Title II of P.L. 108-360 and modeled after NEHRP), and Title III would create an interagency coordinating committee, chaired by the Director of NIST, that would oversee the planning and coordination of both the earthquake and wind hazards programs. The single interagency coordinating committee would replace the two separate interagency committees overseeing the current earthquake and wind hazards programs. The bill was reported by the House Science and Technology Committee on February 26, 2010, and was passed by the House on March 2, 2010. No action on the bill has been taken in the Senate.

The interagency coordinating committee also would be given authority to "make proposals for planning and coordination of any other federal research for natural hazards mitigation that the Committee considers appropriate." The potentially broader mandate for the interagency coordinating committee—to embrace all natural hazards in its deliberations—could reflect an emphasis on natural hazard *mitigation* presented in the bill's "Findings" section. The bill finds that research is needed to better understand institutional, social, behavioral, and economic factors that influence how risk mitigation is implemented, and that a major goal of federally supported natural hazards-related research should be to increase the adoption of hazard mitigation measures. This theme is an aspect of an "all-hazards" approach to reducing risk, and could reflect observations presented during a June 11, 2009, hearing of the House Science and Technology Committee at which a witness noted that "there appear to be significant similarities in societal responses to different hazards."[4] In that testimony, the witness observed that there are limited opportunities to study earthquake emergency response and recovery because damaging earthquakes are relatively infrequent; however, there are important lessons to be learned from studying other, more frequent, hazards, such as tornadoes.[5]

Authorization of Appropriations in H.R. 3820

H.R. 3820 would authorize total appropriations for NEHRP of approximately $906 million for a five-year period ending in FY2014, with 90% of the funding authorized for the USGS and NSF, and the remainder for FEMA and NIST. (See Table 2.) The total authorized amounts would be slightly greater than what was authorized by P.L. 108-360 over five years. Also, funding under H.R. 3820 would exceed the amounts actually appropriated between FY2005 and FY2009 by over 60% for NEHRP if Congress enacted appropriations over five years that matched the authorized amounts. Compared to the previously enacted authorized funding, total authorized amounts would be greater for USGS and NSF under H.R. 3820 and less for NIST and FEMA.

Table 2. NEHRP Authorization for Appropriations Under H.R. 3820 ($ millions)

	FY2010	FY2011	FY2012	FY2013	FY2014	Total FY2010-FY2014 auth.	Total FY2005-FY2009 auth.	Total FY2005-FY2009 enact.
USGS	90.0	92.1	94.3	96.5	98.8	*471.7*	*423.6*	*287.6*
NSF	64.1	66.1	68.0	70.1	72.2	*340.5*	*306.3*	*272.0*
FEMA	10.2	10.6	10.9	11.2	11.5	*54.4*	*111.5*	*46.6*
NIST	7.0	7.7	7.9	8.2	8.4	*39.2*	*61.0*	*9.3*
Total	*171.3*	*176.5*	*181.1*	*186*	*190.9*	*905.8*	*902.4*	*615.5*

Source: U.S. House of Representatives, H.R. 3820; and NEHRP program office, at http://www.nehrp.gov/pdf/ ppt_sdr.pdf.

Note: Totals may not sum due to rounding.

The USGS would receive the largest share—about 52%—of total authorized appropriations under H.R. 3820, as under the previous reauthorization of NEHRP, and the total amount would be approximately $48 million more than the amount authorized for FY2005 through FY2009. As with the previous reauthorization, H.R. 3820 singles out the Advanced National Seismic System (ANSS) to receive a subset of authorized appropriations within the total USGS-authorized amount. Specifically, ANSS would be authorized to receive

$36 million in FY2010, $37 million in FY2011, $38 million in FY2012, $39 million in FY2013, and $40 million in FY2014. That would total $190 million over five years, compared to a total of $174 million over five years in the previous reauthorization.

Other Changes Proposed in H.R. 3820

Section 105 of H.R. 3820 would shift a post-earthquake investigations program from the USGS to be led instead by NIST, in consultation with the other NEHRP agencies, and the program would be organized to study the implications of earthquakes in the areas of responsibility of each NEHRP agency.

The interagency coordinating committee that would be created under Title III of H.R. 3820 would be largely similar to the current interagency committee except that it would also include the National Oceanic and Atmospheric Administration (NOAA), and the committee would have the discretion to include the head of any other federal agency it considers appropriate. In addition to acting as a single coordinating committee for both the earthquake and wind hazards program, the committee is charged with developing a strategic plan for both programs,[6] providing progress reports to Congress, and developing coordinated budgets for each program. Also, the Director of NIST is required to establish advisory committees for both programs, similar to the current advisory committees established under P.L. 108-360.

Title III of H.R. 3820 also requires the Subcommittee on Disaster Reduction of the Committee on Environment and Natural Resources of the National Science and Technology Council to submit a report, within two years of enactment, identifying federal research, development, and technology transfer activities for natural disasters, common areas of research among the natural hazards, and opportunities to "create synergies between the research activities for the hazards."

EARTHQUAKE HAZARDS AND RISK

Figure 2 indicates that detailed information exists on where earthquakes are likely to occur in the United States and how severe the earthquake magnitude and resulting ground shaking are likely to be. The map in Figure 2 depicts the potential shaking hazard from future earthquakes. It is based on the frequency at which earthquakes occur in different areas and how far the strong shaking extends from the source of the earthquake. In Figure 2, the hazard levels indicate the potential ground motion—expressed as a percentage of the acceleration due to gravity (g). In a sense, the map shows the likelihood of where earthquakes could occur, and where the strongest shaking could take place.

All 50 states and the District of Columbia are vulnerable to earthquake hazards, although risks vary greatly across the country. Seismic hazards are greatest in the western United States, particularly California, Washington, Oregon, and Alaska and Hawaii. Alaska is the most earthquake-prone state, experiencing a magnitude 7 earthquake almost every year and a magnitude 8 earthquake every 14 years on average. (See box below for a description of earthquake magnitude.) Because of its low population and infrastructure density, Alaska has a relatively low risk for large economic losses from an earthquake. In contrast, California has more citizens and infrastructure at risk than any other state because of the state's frequent seismic activity combined with its high population.

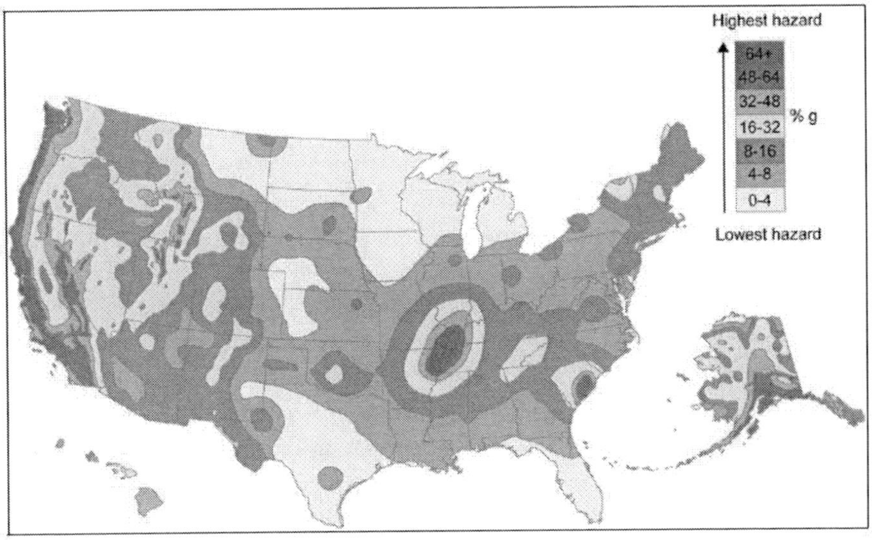

Source: USGS Fact Sheet 2008-3018 (April 2008), at http://earthquake.usgs.gov/research/hazmaps/products_data/images/nshm_us02.gif. Modified by CRS.

Note: The bar in the upper right shows the potential ground motion—expressed as a percentage of the acceleration due to gravity (g)—with up to a 1 in 10 chance of being exceeded over a 50-year period.

Figure 2. Earthquake Hazard in the United States.

Figure 2 also shows relatively high earthquake hazard in the Rocky Mountain region, portions of the eastern seaboard—particularly South Carolina—and a part of the central United States known as the New Madrid Seismic Zone (see "The New Madrid Seismic Zone," below). Other portions of the eastern and northeastern United States are also vulnerable to moderate seismic hazard.

According to the USGS, 75 million people in 39 states are subject to "significant risk."[7]

Earthquake Magnitude and Intensity

Earthquake magnitude is a number that characterizes the relative size of an earthquake. It is often reported using the *Richter* scale (magnitudes in this report are generally consistent with the Richter scale). Richter magnitude is calculated from the strongest seismic wave recorded from the earthquake, and is based on a logarithmic (base 10) scale: for each whole number increase in the Richter scale, the ground motion increases by 10 times. The amount of energy released per whole number increase, however, goes up by a factor of 32. The *moment magnitude* scale is another expression of earthquake size, or energy released during an earthquake, that roughly corresponds to the Richter magnitude and is used by most seismologists because it more accurately describes the size of very large earthquakes. *Intensity* is a measure of how much shaking occurred at a site based on observations and amount of damage. Intensity is usually reported on the Modified Mercalli Intensity Scale as a Roman numeral ranging from I (not felt) to XII (total destruction).

National Seismic Hazards Maps and Earthquake Forecast for California

On April 21, 2008, the USGS released National Seismic Hazards Maps that updated the version published in 2002. Compared to the 2002 version, the new maps indicate lower ground motions (by 10% to 25%) for the central and eastern United States, based on modifications to the ground-motion models used for earthquakes. The new maps indicate that estimates of ground motion for the western United States are as much as 30% lower for certain types of ground motion, called long-period seismic waves, which affect taller, multi-story buildings. Ground motion that affects shorter buildings of a few stories, called short-period seismic waves, is roughly similar to the 2002 maps. The new maps show higher estimates for ground motion for western Oregon and Washington compared to the 2002 maps, due to new ground motion models for the offshore Cascadia subduction zone. In formulating the 2008 maps, the USGS gave more weight to the probability of a catastrophic magnitude 9 earthquake occurring along the Cascadia subduction zone. The Cascadia subduction zone fault ruptures, on average, every 500 years, and has the potential to generate destructive earthquakes and tsunamis along the coasts of Washington, Oregon, and northern California.

According to a report released on April 14, 2008, California has a 99% chance of experiencing a magnitude 6.7 or larger earthquake in the next 30 years. The likelihood of an even larger earthquake, magnitude 7.5 or greater, is 46% and will likely occur in the southern part of the state. The fault with the highest probability of generating at least one earthquake of magnitude 6.7 or greater over the next 30 years is the San Andreas in southern California (59% probability); for northern California it is the Hayward-Rodgers Creek fault (31%). The earthquake forecasts are not predictions (i.e., they do not give a specific date or time), but represent probabilities over a given time period. In addition, the probabilities have variability associated with them. The earthquake forecasts are known as the "Uniform California Earthquake Rupture Forecast (UCERF)" and are produced by a working group composed of the USGS, the California Geological Survey, and the Southern California Earthquake Center.

Sources: USGS Fact Sheet 2008-3018, "2008 United States National Seismic Hazard Maps" (April 2008), at http://pubs.usgs.gov/fs/2008/3018/pdf/FS08-3018_508.pdf; USGS Fact Sheet 2008-3027, "Forecasting California's Earthquakes—What Can We Expect in the Next 30 Years?" (2008), at http://pubs.usgs.gov/fs/2008/3027/fs2008- 3027.pdf.

Shaking hazards maps, such as the one in Figure 2, are often combined with other data, such as the strength of existing buildings, to estimate possible damage in an area due to an earthquake. The combination of seismic risk, population, and vulnerable infrastructure can help improve the understanding of which urban areas across the United States face risks from earthquake hazards that may not be immediately obvious from the probability maps of shaking hazards alone. The USGS has identified 26 urban areas that face a significant seismic risk from the combination of population and severity of shaking. Table 3 lists those areas at greatest risk.

The USGS estimates that several million earthquakes occur worldwide each year, but the majority are of small magnitude or occur in remote areas, and are not detectable. More earthquakes are detected each year as more seismometers[8] are installed in the world, but the number of large earthquakes (magnitude greater than 6.0)[9] has remained relatively constant.

Between 2000 and 2008 there were between 2,261 and 3,876 earthquakes per year in the United States, according to the National Earthquake Information Center (NEIC). (See Figure 3.)

Table 3. Urban Areas Facing Significant Seismic Risk (alphabetically by state for cities with at least 300,000 people)

State	City	State	City
Alaska	Anchorage	Nevada	Las Vegas
California	Fresno	Nevada	Reno
California	Los Angeles	New Mexico	Albuquerque
California	Sacramento	New York	New York
California	Salinas	Oregon	Eugene-Springfield
California	San Diego	Oregon	Portland
California	San Francisco-Oakland	Puerto Rico	San Juan
California	Santa Barbara	South Carolina	Charleston
California	Stockton-Lodi	Tennessee	Chattanooga-Knoxville
Idaho	Boise	Tennessee	Memphis
Indiana	Evansville	Utah	Provo-Orem
Massachusetts	Boston	Utah	Salt Lake City
Missouri	St. Louis	Washington	Seattle

Sources: USGS Fact Sheet 2006-3016 (March 2006); USGS Circular 1188, Table 3.

Note: These areas are identified using a population-based risk factor based on 1999 population data. (William Leith, ANSS Coordinator, USGS, Reston, VA, telephone conversation, Nov. 15, 2006).

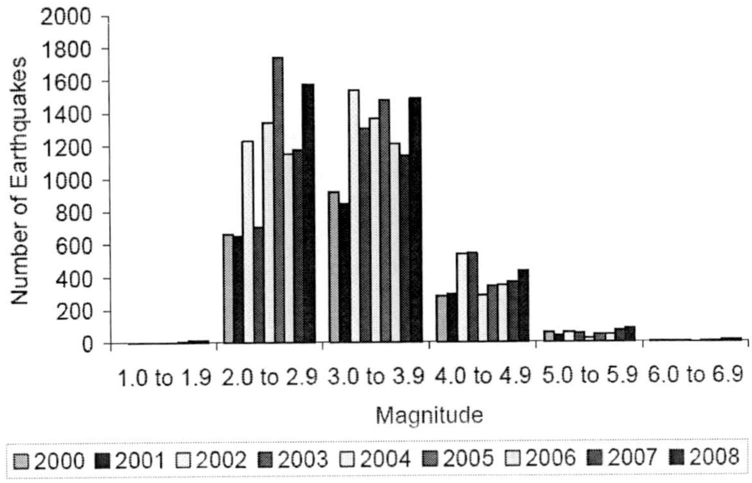

Source: USGS, "Earthquake Facts and Statistics," at http://neic.usgs.gov/neis/eqlists/eqstats.html; data as of December 7, 2009.

Note: Earthquakes greater than magnitude 7.0 and less than 1.0 are not shown. According to the USGS, 6 earthquakes of magnitude 7.0 or greater occurred in the United States between 2001 and 2007.

Figure 3. Histogram of the Number of U.S. Earthquakes from 2000 to 2008 by Magnitude (1.0 to 6.9).

As Figure 3 shows, about 98% of earthquakes detected each year by the NEIC are smaller than magnitude 5.0; only 59 earthquakes exceeded magnitude 6.0 for the nine-year period (less than 0.3% of the total earthquakes detected) for an average of less than seven earthquakes per year of at least 6.0 magnitude. Large earthquakes, although infrequent, cause the most damage and are responsible for most earthquake-related deaths. The great San Francisco earthquake of 1906 claimed an estimated 3,000 lives, as a result of both the earthquake and subsequent fires. Over the past 100 years, relatively few Americans have died as a result of earthquakes, compared to citizens in some other countries.[10] Since 1970, three major earthquakes in the United States were responsible for 188 of the 212 total earthquake-related fatalities (see **Table 4**).

Table 4. Earthquakes Responsible for Most U.S. Fatalities Since 1970

Date	Location	Magnitude	Deaths
February 9, 1971	San Fernando Valley, CA	6.6	65
October 18, 1989	Loma Prieta, CA	6.9	63
January 17, 1994	Northridge, CA	6.7	60

Source: USGS, http://earthquake.usgs.gov/regional/states/us_deaths.php.

Note: Other sources report different numbers of fatalities associated with the Northridge earthquake.

Since 2000, only two deaths directly caused by earthquakes have occurred in the United States, both associated with falling debris in Paso Robles (CA) during the December 22, 2003, San Simeon earthquake of magnitude 6.5. In contrast, earthquakes have been directly or indirectly responsible for more than 685,000 fatalities in other countries since 2000.[11] Approximately 65% of those estimated deaths resulted from the December 2004 Indonesian earthquake (and resulting tsunami) of magnitude 9.1, and the January 2010 magnitude 7.0 earthquake in Haiti. Since the devastating Indonesian earthquake occurred, an October 2005 magnitude 7.6 earthquake in Pakistan killed approximately 86,000 people, and a May 2008 magnitude 7.9 earthquake struck Eastern Sichuan, China, causing as many as 87,000 fatalities (see box).

The 1994 Northridge earthquake was the nation's most damaging earthquake in the past 100 years, preceded five years earlier by the second most costly earthquake—Loma Prieta. **Table 5** shows the 10 costliest U.S. earthquakes in terms of insured and uninsured losses. Comparing losses between different earthquakes, and between earthquakes and other disasters such as hurricanes, can be difficult because of the different ways losses are calculated. Calculations may include a combination of insured losses, uninsured losses, and estimates of lost economic activity.

For example, insured losses from Hurricane Katrina in 2005—mainly property—may be $41 billion, according to one estimate.[12] Total property damage would rise if uninsured property were included; and including interrupted economic activity in the calculation could bring the total loss for Hurricane Katrina to $100 billion, according to one estimate.[13]

The May 12, 2008, Sichuan Earthquake in China and Implications for the United States

On May 12, 2008, a catastrophic earthquake of magnitude 7.9 struck Eastern Sichuan, China. The epicenter was located approximately 960 miles southwest of Beijing, and the earthquake was triggered approximately 12 miles below the earth's surface. Approximately 87,000 fatalities have been reported. The earthquake was felt in parts of eastern, southern, and central China, and as far away as Bangladesh, Taiwan, Thailand, and Vietnam. Several large aftershocks occurred after the main seismic event.

The May 12 earthquake resulted from movement along a northeast-trending *reverse* or *thrust* fault, reflecting stresses from the convergence of rocks of the Tibetan Plateau, to the west, against the crust underlying the Sichuan Basin and southeastern China. The region has experienced large earthquakes in the past; on August 25, 1933, a magnitude 7.5 earthquake struck the northwestern margin of the Sichuan Basin, resulting in approximately 9,300 fatalities.

Some concerns have been raised about the possibility of an earthquake of similar magnitude occurring in a seismically active region of the United States, such as southern California, where fault movement similar to the Eastern Sichuan earthquake may occur. On May 22, 2008, the USGS released a hypothetical scenario for a magnitude 7.8 southern California earthquake, called the ShakeOut Scenario. In the scenario, scientists hypothetically simulated the ground shaking and fault rupture associated with a magnitude 7.8 earthquake, and estimated the resulting damage to buildings and infrastructure. The scenario estimated approximately 1,800 fatalities and $213 billion in economic losses as a result of the earthquake. The report points to aggressive retrofitting programs that have increased the seismic resistance of buildings, highways, and other critical infrastructure in southern California as one reason why the number of possible fatalities is relatively low.

Some scientists have raised the possibility that earthquakes, such as the May 12, 2008, Sichuan event, may sometimes exhibit cascading behavior, where bursts of seismic energy are released along different places in a single fault, or jump between connected faults. Earthquakes that occur along the Sierra Madre fault in southern California, for example, could trigger a series of cascading seismic events along other faults, such as the San Andreas. Seismic hazard estimates may not fully account for the damage that could be caused by cascading earthquakes along a connected fault system. Scientists are hoping to examine the Sichuan earthquake in more detail to better understand the nature of cascading seismic events and how they affect the U.S. seismic hazard estimates.

Sources: Ken Hudnut, geophysicist, USGS, Pasadena, CA, phone conversation, June 11, 2008; USGS Earthquake Hazards Program, at http://earthquake.usgs.gov/eqcenter/eqinthenews/2008/us2008ryan/#summary; USGS Newsroom, *Earthquake Fatalities High in 2008*, at http://www.usgs.gov/newsroom/article.asp?ID=2101; and USGS, The ShakeOut Scenario, Open-File Report 2008-1150 (2008), at http://pubs.usgs.gov/of/2008/1150/.

Table 5. The 10 Most Damaging Earthquakes in the United States

Year	Location	Magnitude	$2005 (billions)
1994	Northridge, CAa	6.7	$26.0
1989	Loma Prieta, CA	6.9	$11.0
1964	Anchorage, AK	9.2	$3.1
1971	San Fernando, CA	6.5	$2.7
2001	Nisqually, WA	6.8	$2.5
1987	Whittier Narrows, CA	5.9	$0.62
1933	Long Beach, CA	6.3	$0.60
1953	Kern County, CA	7.5	$0.44
1992	Landers, CA	7.6	$0.13
1992	Cape Mendocino, CA	7.1	$0.092

Source: Insurance Information Institute, at http://www.iii.org/media/facts/statsbyissue/earthquakes/.
Note: Includes insured and uninsured losses.
a. Estimates for total losses resulting from the Northridge earthquake vary; the Congressional Budget Office estimated $43 billion in total losses ($50 million in 2005 dollars). See *Federal Reinsurance for Disasters*, Congressional Budget Office (September 2002), p. 19.

January 12, 2010, Magnitude 7.0 Earthquake Strikes Haiti

On Tuesday, January 12, 2010, a magnitude 7.0 earthquake struck Haiti at 4:53 p.m. The epicenter was located approximately 15 miles southwest of Port-au-Prince, and the earthquake occurred at a depth of about 8 miles, according to initial USGS reports. The relatively shallow earthquake, and its close proximity to the capital city, exposed millions of Haitians to severe to violent ground shaking. The earthquake occurred along the Enriquillo-Plantain Garden fault system, a major east-west trending strike-slip fault system that lies between the Caribbean tectonic plate and the North American tectonic plate; the Caribbean plate actively moves against the North American plate and shear stresses are created at the boundary. At a strike-slip fault, the rocks move past each other horizontally along the fault line (in contrast to a thrust fault, where rocks on one side of the fault move on top of the rocks on the other side). Earthquakes comparable to or stronger than the January 12 event have struck Haiti at least four times over the past three centuries. Earthquakes in 1751 and 1770—which probably occurred along the same fault system—destroyed Port-au-Prince. Other examples of strike-slip faults are the San Andreas fault in California and the Red River fault in China.

The January 12 earthquake caused widespread damage in the Port-au-Prince area, causing approximately 230,000 deaths and 300,000 injuries. Also, a series of aftershocks followed the main earthquake. There were 14 aftershocks greater than magnitude 5 and 36 greater than magnitude 4 within the first day following the magnitude 7.0 event. Aftershocks have the potential to cause further damage, especially to structures weakened by the initial large earthquake. In a February 23, 2010, statement, the USGS indicated that the frequency of aftershocks should diminish with time, but the threat of additional damaging earthquakes remains.

The USGS noted that buildings in the Port-au-Prince area will continue to be at risk from strong earthquake shaking, and that the fault responsible for the January 12 earthquake still stores sufficient strain to be released as a large, damaging earthquake during the lifetime of structures built during the reconstruction effort.

The USGS based its probability estimates on techniques developed to assess earthquake hazards in the United States. Using these techniques, the USGS estimated that the probability of a magnitude 7 or greater earthquake occurring within the next 50 years along the Enriquillo fault near Port-au-Prince is between 5% and 15%. The range of probabilities reflects the current understanding of the seismicity and tectonics of the Haiti region. By comparison, the USGS has estimated that that the probability of a magnitude 7 or greater earthquake occurring within the next 50 years along the Hayward-Rodgers Creek fault east of San Francisco is about 15%.

Sources: USGS Earthquake Hazards Program, Significant Earthquakes: Magnitude 7.0—Haiti Region, at http://earthquake.usgs.gov/ earthquakes/ eqinthenews/2010/us2010rja6/; Michael Blanpied, Associate Coordinator for the USGS Earthquake Hazards Program, podcast, Jan. 13, 2010, at http://www.usgs.gov/corecast/details.asp?ep=117; email from David Applegate, Senior Science Advisor for Earthquake and Geologic Hazards, Jan. 13, 2010; USGS statement, "USGS Updates Assessment of Earthquake Hazard and Safety in Haiti and the Caribbean," February 23, 2010, at http://www.usgs.gov/newsroom/ article.asp?ID=2413&from=rss_home. See also CRS Report R41023, *Haiti Earthquake: Crisis and Response*, by Rhoda Margesson and Maureen Taft-Morales.

Potential Losses from Earthquakes

The United States faces potentially large total losses due to earthquake-caused damage to buildings and infrastructure and lost economic activity. As urban development continues in earthquake-prone regions in the United States, concerns are increasing about the exposure of the built environment, including utilities and transportation systems, to potential earthquake damage.14 One estimate of loss from a severe earthquake in the Los Angeles area is over $500 billion. An even higher estimate—approximately $900 billion—includes damage to the heavily populated central New Jersey-Philadelphia corridor if a 6.5 magnitude earthquake occurred along a fault lying between New York City and Philadelphia.[15]

Some studies and techniques combine seismic risk with the value of the building inventory16 and income losses (e.g., business interruption, wage, and rental income losses) in cities, counties, or regions across the country to provide estimations of economic losses from earthquakes. An April 2008 report from FEMA calculated that the *annualized* loss from earthquakes nationwide is $5.3 billion, with California, Oregon, and Washington accounting for nearly $4.1 billion (77%) of the U.S. total estimated annualized loss.17 **Table 6** shows metropolitan areas with estimated annualized U.S. earthquake losses over $10 million.

Table 6. U.S. Metropolitan Areas with Estimated Annualized Earthquake Losses of More Than $10 Million (in $ millions)

Rank	Metro area	AEL	Rank	Metro area	AEL
1	Los Angeles-Long Beach-Santa Ana, CA	$1,312	23	Reno-Sparks, NV	$29
2	San Francisco-Oakland-Fremont, CA	$781	24	Charleston-North Charleston, SC	$22
3	Riverside-San Bernadino-Ontario, CA	$397	25	Columbia, SC	$22
4	San Jose-Sunnyvale-Santa Clara, CA	$277	26	Stockton, CA	$21
5	Seattle-Tacoma, WA	$244	27	Atlanta-Sandy Springs-Marietta, GA	$19
6	San Diego-Carlsbad-San Marcos, CA	$155	28	Bremerton-Silverdale, WA	$18
7	Portland-Vancouver-Carlsbad, OR	$137	29	Ogden-Clearfield, UT	$18
8	Oxnard-Thousand Oaks-Ventura, CA	$111	30	Salem, OR	$17
9	Santa Rosa-Petaluma, CA	$69	31	Eugene-Springfield, OR	$17
10	St. Louis, MO-IL	$59	32	Napa, CA	$16
11	Salt Lake City, UT	$52	33	San Luis Obispo-Paso Robles, CA	$16
12	Sacramento-Arden-Arcade-Roseville, CA	$52	34	Nashville-Davidson-Murfreesboro, TN	$15
13	Vallejo-Fairfield, CA	$40	35	Albuquerque, NM	$15
14	Memphis, TN	$38	36	Olympia, WA	$14
15	Santa Cruz-Watsonville, CA	$36	37	Modesto, CA	$13
16	Anchorage, AK	$35	38	Fresno, CA	$13
17	Santa Barbara-Santa Maria-Goleta, CA	$34	39	Evansville, IN-KY	$12
18	Las Vegas-Paradise, NV	$33	40	Birmingham-Hoover, AL	$11
19	Honolulu, HI	$32	41	El Centro, CA	$11
20	Bakersfield, CA	$30	42	Little Rock-North Little Rock, AR	$11
21	New York-Northern New Jersey-	$30	43	Provo-Orem, UT	$10
	Long Island, NY				
22	Salinas, CA	$29			

Source FEMA Publication 366, *HAZUS MH Estimated Annualized Earthquake Losses for the United States* (April 2008). Annualized earthquake losses (AEL) calculated in 2005 dollars.

Annualized earthquake loss (AEL) addresses two components of seismic risk: the probability of ground motion and the consequences of ground motion. It enables comparison between different regions with different seismic hazards and different building construction types and quality. For example, earthquake hazard is higher in the Los Angeles area than in Memphis, but the general building stock in Los Angeles is more resistant to the effects of earthquakes. The AEL annualizes the expected losses by averaging them by year.

February 27, 2010, Magnitude 8.8 Earthquake in Chile: Is There a Similar Risk to the United States?

A massive magnitude 8.8 earthquake struck Chile on February 27, 2010, along a subduction zone plate boundary fault 70 miles north-northeast of the city of Concepcion and offshore of the Chilean coast. The earthquake occurred at a depth of approximately 22 miles below the seafloor, much deeper than the earthquake that struck Haiti near Port-au-Prince on January 12. Several cities, including Concepcion, experienced intensity VIII shaking on the Modified Mercalli Intensity Index, corresponding to severe perceived shaking. The capital city of Santiago, located 200 miles northeast of the epicenter, experienced intensity VII shaking corresponding to very strong perceived shaking. The earthquake has caused widespread damage that may amount to $3-$8 billion in insured losses, and probably more in total economic losses. Over 700 deaths have been reported, many from the tsunami generated by the subsea earthquake, and possibly 2 million people have been affected. Far fewer fatalities are expected in Chile compared to the Haiti earthquake, even though the Chilean earthquake was much more powerful, in part because the epicenter was farther from major population centers and the earthquake occurred 14 miles deeper than the Haitian event. In addition, details are beginning to emerge about the effectiveness of earthquake-resistant design and building construction in Chile as compared to Haiti.

Because the earthquake occurred offshore, it generated a tsunami, which struck parts of the Chilean coastline and offshore islands, causing damage and fatalities. Tsunami warnings were issued by the National Weather Service Pacific Tsunami Warning Center for Hawaii, Japan, and other regions bordering the Pacific Ocean that may have been vulnerable to a damaging tsunami wave, although most regions far from the epicenter did not experience any serious damage. A tsunami caused significant damage to the city of Hilo, Hawaii, following the May 1960 magnitude 9.5 earthquake that also occurred along the subduction zone fault about 143 miles south of the February 27, 2010, earthquake. Why the 1960 earthquake generated a tsunami that caused damage and fatalities in Hawaii, Japan, and the Philippines while the 2010 earthquake did not is not yet well understood and is being actively studied.

The magnitude 8.8 earthquake occurred along the boundary between the Nazca tectonic plate and the South American tectonic plate, which converge at a rate of about 3 inches per year. The Nazca plate is subducting under the South American plate, which rides over the top of the Nazca plate. In geologic terms, this is known as a thrust fault or megathrust, in contrast to a strike-slip fault, where the rocks on either side of the fault slide past each other. The San Andreas fault and the Enriquillo fault that caused the January 2010 Haiti earthquake are strike-slip faults. The Sumatran-Andaman megathrust fault, which triggered the December 2004 Indonesian earthquake and tsunami, is a subduction zone fault or megathrust geologically similar to the Nazca-South American tectonic plate subduction zone.

Subduction zone megathrust faults generate the largest earthquakes in the world. The Cascadia Subduction Zone megathrust that stretches from mid-Vancouver Island in southern British Columbia southward to Cape Mendocino in northern California has the potential to generate a very large earthquake, similar in magnitude to the February 2010

Chilean earthquake. The fault's proximity to the northwestern U.S. coastline—approximately 50-100 miles offshore—also poses a significant tsunami hazard; destructive waves from a large earthquake along the fault could reach the coast of Oregon and Washington in less than an hour, possibly in tens of minutes. The Cascadia Subduction Zone fault forms the boundary between the subducting Juan de Fuca tectonic plate and the overriding North American plate, very similar to the relationship between the Nazca Plate and the South American Plate off the Chilean coast. If the Cascadia Subduction Zone megathrust were to "unzip" or rupture along a large section of its entire length, it would likely generate a megathrust earthquake near magnitude 9 or more, similar to the 1964 Alaskan earthquake, the 1960 and 2010 Chilean earthquakes, and the 2004 Indonesian earthquake. Scientists have documented that the last time this occurred along the Cascadia Subduction Zone fault was in 1700. The 1700 earthquake spawned a tsunami that traveled across the Pacific Ocean and struck Japan. Because of the tectonic similarities between the Cascadia Subduction Zone megathrust and the Nazca-South American plate megathrust, scientists hope to learn a great deal about the seismic hazard in the Pacific Northwest by studying the unique strong ground motion recordings from the 2010 Chilean magnitude 8.8 earthquake.

Sources: USGS Earthquake Hazards Program, *Magnitude 8.8–Offshore Maule, Chile*, at http://earthquake.usgs.gov/earthquakes/eqinthenews/2010/us2010tfan/; Eqecat, "M8.8 Earthquake Offshore Maule, Chile," CatWatch, February 27, 2010, at http://www.eqecat.com/catWatchREV/secureSite/report.cfm?id=220 (Eqecat provides loss estimates for insurance and financial institutions); *Business Week*, "Chile Asks for International Aid as Looting Spreads," March 2, 2010, at http://www.businessweek.com/news/2010-03-02/chile-asks-for-international-aid-as-looting-spreads-correct-.html; Brian F. Atwater et al., *The Orphan Tsunami of 1700—Japanese Clues to a Parent Earthquake in North America*, USGS, Professional Paper 1707, 2005, http://pubs.usgs.gov/pp/pp1707/; e-mail from Richard Aster, Professor of Geophysics, New Mexico Institute of Mining and Technology, March 2, 2010.

A Decrease in Estimated Loss?

In its most recent publication estimating earthquake losses, FEMA noted that the $5.3 billion in annualized earthquake loss nationwide was 21% higher than the $4.4 billion calculated in FEMA's previous report, published in February 2001.18 However, the 2001 report calculated losses using 1994 dollars, and when adjusted to reflect 2005 dollars the earlier estimate increased to $5.6 billion, indicating a small decrease in nationwide annualized earthquake loss potential since the 2001 report was published. According to FEMA, this loss occurred even though the national building inventory increased by 50% over this same period.

What factors led to a decreased estimate in potential loss despite growth in building inventory? According to FEMA, two primary factors were responsible: (1) a slight decrease in estimated earthquake hazard in the western United States (namely California) except for some parts of Washington and Utah, and (2) a change in the distribution of building inventory in California, with an increase in wood frame buildings of 17% and a reduction in the amount of masonry (-6%), steel (-5.8%), and concrete (-3%) buildings in the state.19 Wood frame buildings are less vulnerable to earthquake damage, generally, compared to other construction types. Because California accounts for 66% of the overall nationwide annualized earthquake loss, a 17% increase in wood frame buildings had a proportionally large effect. In fact, FEMA attributed 78% of the loss reduction between 2001 and 2008 to the change in building inventory distribution, and 22% to the decrease in earthquake hazard for California.20

A single large earthquake can cause far more damage than the average annual estimate. Annualized estimates, however, help provide comparisons of infrequent, high impact events like damaging earthquakes, with more frequently occurring hazards like floods, hurricanes, or other types of severe weather. The annualized earthquake loss values shown in Table 6 represent future estimates, and are calculated by multiplying losses from potential future ground motions by their respective frequencies of occurrence, and then summing these values.21

Table 6 also shows that annualized earthquake losses in the New York-Northern New Jersey-Long Island metropolitan area are $30 million (ranked 21 out of 43 metropolitan areas with losses greater than $1 million per year), even though no destructive earthquakes have struck that area for generations.22 This area has a relatively low seismic hazard, but also has an extensive infrastructure and is densely populated. That combination of seismic risk, extensive infrastructure, and dense population produces a significant risk to people and structures, according to some estimates.23 In the absence of any significant or damaging earthquakes for that region in recent memory, however, the actual risk may be difficult to grasp intuitively.

The New Madrid Seismic Zone

The New Madrid Seismic Zone in the central United States is vulnerable to large but infrequent earthquakes. A series of large (magnitude greater than 7.0) earthquakes struck the Mississippi Valley over the winter of 1811-1812, centered close to the town of New Madrid, MO. Some of the tremors were felt as far away as Charleston, SC, and Washington, DC. The mechanism for the earthquakes in the New Madrid zone is poorly understood,24 and no earthquakes of comparable magnitude have occurred in the area since these events.

Estimating earthquake damage is not an exact science and depends on many factors. As described above, these are primarily the probability of ground motion occurring in a particular area (see Figure 2), and the consequences of that ground motion, which are largely a function of building construction type and quality, and of the level of ground motion and shaking during the actual event. Such factors contribute to the difficulty of making a reasonable damage estimate for a lowfrequency, high-impact event in the New Madrid region based on the probability of an earthquake of similar magnitude occurring. This uncertainty has implications for policy decisions to ameliorate risk, such as setting building codes, and for designing and building structures to withstand a level of shaking commensurate with the risk. Developers of building codes tend to err on the side of caution; presumably the same seismic hazard should lead to similar building codes in urban areas (e.g., compare the seismic hazard for the New Madrid area with parts of California shown in Figure 2).

Some researchers have questioned whether erring on the side of caution in the New Madrid Seismic Zone is justified.25 These researchers question whether the benefits of building structures to conform with the earthquake probability estimates merit the costs, in light of the uncertainty in making those probability estimates.26 These analyses may call into question whether the probability of ground motion estimates for the New Madrid Seismic

Zone (the bulls-eye-shaped area shown in Figure 2 that includes parts of Arkansas, Illinois, Tennessee, and Missouri), and other regions of the country that experience infrequent earthquakes, are too high.27 An uncertainty analysis of the seismic hazard in the New Madrid Seismic Zone is beyond the scope of this report.

MONITORING

Congress authorized the USGS to monitor seismic activity in the United States in the 1990 reauthorization of the National Earthquake Hazards Reduction Act (P.L. 101-614). The USGS operates a nationwide network of seismographic stations called the Advanced National Seismic System (ANSS), which includes the National Strong-Motion Project (NSMP). Globally, the USGS and the Incorporated Research Institutions for Seismology (IRIS) operate 140 seismic stations of the Global Seismic Network (GSN) in more than 80 countries. The GSN provides worldwide coverage of earthquakes, including reporting and research.28

Advanced National Seismic System (ANSS)

According to the USGS, "the mission of ANSS is to provide accurate and timely data and information products for seismic events, including their effects on buildings and structures, employing modern monitoring methods and technologies."[29] If fully implemented, ANSS would encompass more than 7,000 earthquake sensor systems covering parts of the nation vulnerable to earthquake hazards. As envisioned, the system would consist of dense urban networks, regional networks, and backbone stations.

ANSS Funding

Congress first authorized the ANSS program in P.L. 106-503 at a level of $38 million for FY2002 and $44 million for FY2003. The 2004 reauthorization of NEHRP (P.L. 108-360) authorized $30 million for ANSS in FY2005 and then $36 million per year through FY2009. The USGS spent $1.6 million in FY2000 and $3.6 million in FY2001 on ANSS-directed funding, but expenditures have never reached authorized levels since Congress first authorized appropriations for ANSS. From FY2005 through FY2009, the USGS spent a total of approximately $42.5 million within its Earthquake Hazards Program on ANSS, or approximately 24% of the total authorized levels over the five-year period since NEHRP was last reauthorized.30

The FY2010 budget request stated that the USGS plans to install a cumulative total of 822 ANSS monitoring stations by the end of 2009. That would represent approximately 12% of the 7,000 seismic stations originally envisioned for the program. According to its budget justification, the USGS plans to devote its ANSS-directed resources to operating and maintaining the installed system.31 Of the approximately $8.8 million for ANSS-directed funding in FY2009, about $1.5 million was devoted to the development, modernization, and expansion of the system; the remainder of FY2009 funding was used to operate the existing system.32 However, the American Recovery and Reinvestment Act (ARRA, P.L. 111-5) provided approximately $19 million in ANSS-directed funding to be spent over FY2009-

FY2011.33 All of the ARRA funding would be provided for modernization and expansion of the current system, and when supplemented by base program funds over the three-year period, would double the number of ANSS stations.34 A doubling of the number of current stations would total approximately 23% of the 7,000 stations originally planned for ANSS.

Dense Urban Networks

In the original conception for ANSS, approximately 6,000 of the planned stations would be installed in 26 high-risk urban areas to monitor strong ground shaking and how buildings and other structures respond. Currently, five high-risk urban areas have instruments deployed in sufficient density to generate the data to produce near real-time maps, called ShakeMaps, which can be used in emergency response during and after an earthquake.35 (See "ShakeMap," below.)

Regional Networks

Approximately 1,000 new instruments would replace aging and obsolete stations in the networks that now monitor the nation's most seismically active regions. The current regional networks contain a mix of modern, digital, broadband, and high-resolution instruments that can provide real-time data; they are supplemented by older instruments that may require manual downloading of data. Universities in the region typically operate the regional networks and will likely continue to do so as ANSS is implemented.

Backbone Stations

Approximately 100 instruments comprise the existing "backbone" of ANSS, with a roughly uniform distribution across the United States, including Alaska and Hawaii. These instruments provide a broad and uniform minimum threshold of coverage across the country. The backbone network consists of USGS-deployed instruments and other instruments that serve both ANSS and the EarthScope project (described below, under "National Science Foundation").

National Strong-Motion Project (NSMP)

Under ANSS, the USGS operates the NSMP to record seismic data from damaging earthquakes in the United States on the ground and in buildings and other structures in densely urbanized areas. The program currently has 900 strong-motion[36] instruments in 701 permanent stations across the United States and in the Caribbean. The NSMP has three components: data acquisition, data management, and research. The near real-time measurements collected by the NSMP are used by other government agencies for emergency response and real-time warnings. If fully implemented, the ANSS program would deploy about 3,000 strong-motion instruments. Many of the current NSMP instruments are older designs and are being upgraded with modern seismometers.

Global Seismic Network (GSN)

The GSN is a system of broadband digital seismographs arrayed around the globe and designed to collect high-quality data that are readily accessible to users worldwide, typically via computer. Currently, 140 stations have been installed in 80 countries and the system is nearly complete, although in some regions the spacing and location of stations has not fully met the original goal of uniform spacing of approximately 2,000 kilometers. The system is currently providing data to the United States and other countries and institutions for earthquake reporting and research, as well as for monitoring nuclear explosions to assess compliance with the Comprehensive Test Ban Treaty. Funding for the GSN totaled approximately $9 million in FY2009.[37]

The Incorporated Research Institutions for Seismology (IRIS)[38] coordinates the GSN and manages and makes available the large amounts of data that are generated from the network. The actual network of seismographs is organized into two main components, each managed separately. The USGS operates two-thirds of the stations from its Albuquerque Seismological Laboratory, and the University of California-San Diego manages the other third via its Project IDA (International Deployment of Accelerometers). Other universities and affiliated agencies and institutions operate a small number of additional stations. IRIS, with funding from the NSF, supports all of the stations not funded through the USGS appropriations.

DETECTION, NOTIFICATION, AND WARNING

Unlike other natural hazards, such as hurricanes, where predicting the location and timing of landfall is becoming increasingly accurate, the scientific understanding of earthquakes does not yet allow for precise earthquake prediction. Instead, notification and warning typically involves communicating the location and magnitude of an earthquake as soon as possible after the event to emergency response providers and others who need the information.

Some probabilistic earthquake forecasts are now available that give, for example, a 24-hour probability of earthquake aftershocks for a particular region, such as California. These forecasts are not predictions, and are currently intended to increase public awareness of the seismic hazard, improve emergency response, and increase scientific understanding of the short-term hazard.[39] In the California example, a time-dependent map is created and updated every hour by a system that considers all earthquakes, large and small, detected by the California Integrated Seismic Network,[40] and calculates a probability that each earthquake will be followed by an aftershock[41] that can cause strong shaking. The probabilities are calculated from known behavior of aftershocks and the possible shaking pattern based on historical data.

When a destructive earthquake occurs in the United States or in other countries, the first reports of its location, or epicenter,42 and magnitude originate either from the National Earthquake Information Center (NEIC), or from one of the regional seismic networks that are part of ANSS. Other organizations, such as universities, consortia, and individual seismologists may also contribute information about the earthquake after the event. Products such as ShakeMap are assembled as rapidly as possible to assist in emergency response and damage estimation following a destructive earthquake.

National Earthquake Information Center (NEIC)

The NEIC, part of the USGS, is located in Golden, CO. Originally established as part of the National Ocean Survey (U.S. Department of Commerce) in 1966, the NEIC was made part of the USGS in 1973. With data gathered from the networks described above and from other sources, the NEIC determines the location and size of all destructive earthquakes that occur worldwide and disseminates the information to the appropriate national or international agencies, government public information channels, news media, scientists and scientific groups, and the general public.

With the advent of the USGS Earthquake Notification Service (ENS), notifications of earthquakes detected by the ANSS/NEIC are provided free to interested parties. Users of the service can specify the regions of interest, establish notification thresholds of earthquake magnitude, designate whether they wish to receive notification of aftershocks, and even set different magnitude thresholds for daytime or nighttime to trigger a notification.

The NEIC has long-standing agreements with key emergency response groups, federal, state, and local authorities, and other key organizations in earthquake-prone regions who receive automated alerts—typically location and magnitude of an earthquake—within a few minutes of an event in the United States. The NEIC sends these preliminary alerts by email and pager immediately after an earthquake's magnitude and epicenter are automatically determined by computer.43 This initial determination is then checked by around-the-clock staff who confirm and update the magnitude and location data.44 After the confirmation, a second set of notifications and confirmations are triggered to key recipients by email, pager, fax, and telephone.

For earthquakes outside the United States, the NEIC notifies the State Department Operations Center, and often sends alerts directly to staff at American embassies and consulates in the affected countries, to the International Red Cross, the U.N. Department of Humanitarian Affairs, and other recipients who have made arrangements to receive alerts.

ShakeMap

Traditionally, the information commonly available following a destructive earthquake has been epicenter and magnitude, as in the data provided by the NEIC described above. Those two parameters by themselves, however, do not always indicate the intensity of shaking and extent of damage following a major earthquake. Recently, the USGS developed a product called ShakeMap that provides a nearly real-time map of ground motion and shaking intensity following an earthquake in areas of the United States where the ShakeMap system is in place. Figure 4 shows an example of a ShakeMap.

The maps produced portray the extent of damaging shaking and can be used by emergency response and for estimating loss following a major earthquake. Currently, ShakeMaps are available for northern California, southern California, the Pacific Northwest, Nevada, Utah, Hawaii, and Alaska.45

With improvements to the regional seismographic networks in the areas where ShakeMap is available, new real-time telemetry from the region, and advances in digital communication and computation, ShakeMaps are now triggered automatically and made available within minutes of the event via the Web. In addition, better maps are now available because of recent improvements in understanding the relationship between the ground motions recorded during the earthquake and the intensity of resulting damage. If databases containing inventories of buildings and lifelines are available, they can be combined with shaking intensity data to produce maps of estimated damage. The ShakeMaps have limitations, especially during the first few minutes following an earthquake before more data arrive from distributed sources. Because they are generated automatically, the initial maps are preliminary, and may not have been reviewed by experts when first made available. They are considered a work in progress, but are deemed to be very promising, especially as more modern seismic instruments are added to the regional networks under ANSS and computational and telecommunication abilities improve.

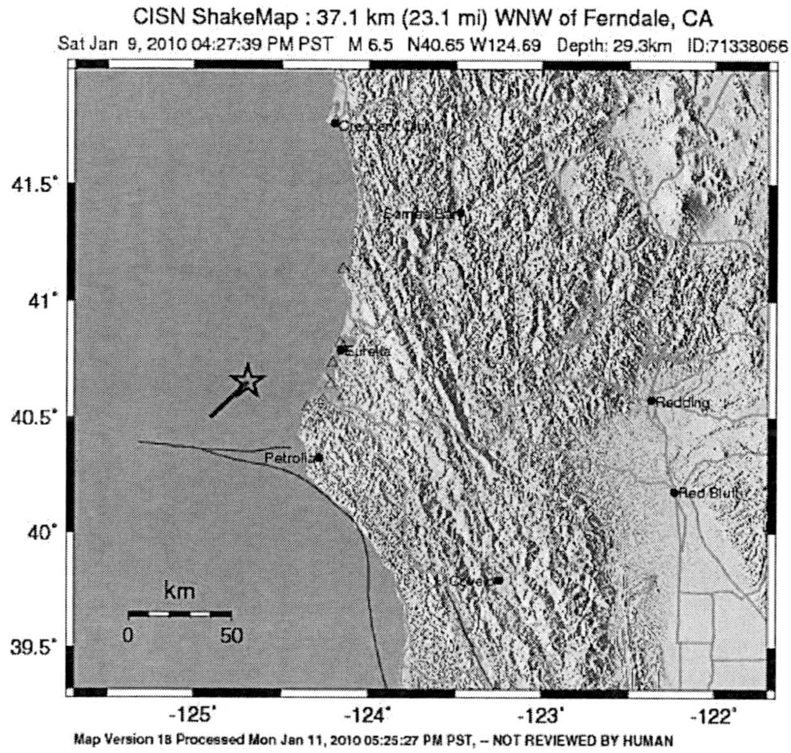

PERCEIVED SHAKING	Not felt	Weak	Light	Moderate	Strong	Very strong	Severe	Violent	Extreme
POTENTIAL DAMAGE	none	none	none	Very light	Light	Moderate	Moderate/Heavy	Heavy	Very Heavy
PEAK ACC.(%g)	<.17	.17-1.4	1.4-3.9	3.9-9.2	9.2-18	18-34	34-65	65-124	>124
PEAK VEL.(cm/s)	<0.1	0.1-1.1	1.1-3.4	3.4-8.1	8.1-16	16-31	31-60	60-116	>116
INSTRUMENTAL INTENSITY	I	II-III	IV	V	VI	VII	VIII	IX	X+

Source: USGS, http://earthquake.usgs.gov/eqcenter/shakemap/nc/shake/71338066/.

Note: Earthquake occurred 23.1 miles west-northwest of Ferndale, CA, at 4:27 p.m. on January 9, 2010, with a magnitude of 6.5. The star indicates the epicenter of the earthquake. Viewed on January 12, 2010. According to preliminary news reports, some damage estimates are nearly $30 million for the town of Eureka itself from the earthquake. (See http://www.times-standard.com/news/ci_14171082.)

Figure 4. Example of a ShakeMap.

Prompt Assessment of Global Earthquakes for Response (PAGER)

Another USGS product that is designed to provide nearly real-time earthquake information to emergency responders, government agencies, and the media is the Prompt Assessment of Global Earthquakes for Response, or PAGER, system.46 This automated system rapidly assesses the number of people, cities, and regions exposed to severe shaking by an earthquake, and generally makes results available within 30 minutes. Following the determination of earthquake location and magnitude, the PAGER system calculates the degree of ground shaking using the methodology developed for ShakeMap, estimates the number of people exposed to various levels of shaking, and produces a description of the vulnerability of the exposed population and infrastructure. The vulnerability includes potential for earthquake-triggered landslides, which could be devastating, as was the case for the huge May 12, 2008, earthquake in Sichuan, China.

The automated and rapid reports produced by the PAGER system provide an advantage compared to the traditional accounts from eye-witnesses on the ground or media reports, because communications networks may have been disabled from the earthquake. Emergency responders, relief organizations, and government agencies could make plans based on PAGER system reports even before getting "ground-truth" information from eye-witnesses and the media.[47]

Figure 5 shows an example of PAGER output for the January 12, 2010, magnitude 7.0 earthquake in Haiti.

Pre-disaster Planning: HAZUS-MH

FEMA developed a methodology and software program called the Hazards U.S. Multi-Hazard (HAZUS-MH).48 The program allows a user to estimate losses from damaging earthquakes, hurricane winds, and floods before a disaster occurs. The pre-disaster estimates could provide a basis for developing mitigation plans and policies, preparing for emergencies, and planning response and recovery. HAZUS-MH combines existing scientific knowledge about earthquakes (for example, ShakeMaps, described above), engineering information that includes data on how structures respond to shaking, and geographic information system (GIS) software to produce maps and display hazards data including economic loss estimates. The loss estimates produced by HAZUS-MH include

- physical damage to residential and commercial buildings, schools, critical facilities, and infrastructure;
- economic loss, including lost jobs, business interruptions, repair and reconstruction costs; and
- social impacts, including estimates of shelter requirements, displaced households, and number of people exposed to the disaster.

In addition to furnishing information as part of earthquake mitigation efforts, HAZUS-MH can also be used to support real-time emergency response activities by state and federal agencies after a disaster. Twenty-seven HAZUS-MH user groups—cooperative ventures among private, public, and academic organizations that use the HAZUS-MH software—have formed across the United States to help foster better-informed risk management for earthquakes and other natural hazards.[49]

HAZUS-MH software was first released to the public in 1997 and the first user group, the Bay Area HAZUS-MH User Group, was formed the same year.

USGS
science for a changing world

USAID FROM THE AMERICAN PEOPLE

M 7.0, HAITI REGION

Origin Time: Tue 2010-01-12 21:53:10 UTC
Location: 18.46°N 72.53°W Depth: 13 km

PAGER
Version 7
Created: 1 day, 4 hours after earthquake

Estimated Population Exposed to Earthquake Shaking

ESTIMATED POPULATION EXPOSURE (k = x1000)		--*	--*	5,887k*	7,261k	1,049k	571k	314k	2,246k	332k
ESTIMATED MODIFIED MERCALLI INTENSITY		I	II-III	IV	V	VI	VII	VIII	IX	X+
PERCEIVED SHAKING		Not felt	Weak	Light	Moderate	Strong	Very Strong	Severe	Violent	Extreme
POTENTIAL DAMAGE	Resistant Structures	none	none	none	V. Light	Light	Moderate	Moderate/Heavy	Heavy	V. Heavy
	Vulnerable Structures	none	none	none	Light	Moderate	Moderate/Heavy	Heavy	V. Heavy	V. Heavy

*Estimated exposure only includes population within the map area.

Population Exposure

population per ~1 sq. km from Landscan

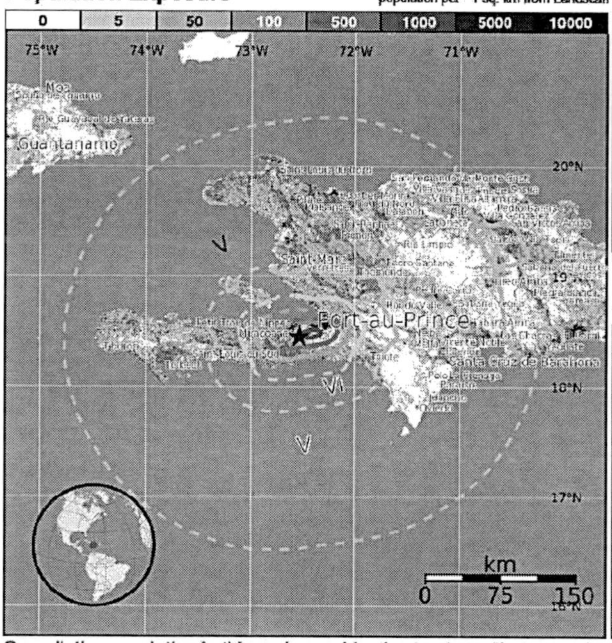

Selected City Exposure

MMI	City	Population
X	Grand Goave	5k
IX	**Port-au-Prince**	1,235k
IX	Carrefour	442k
IX	Petionville	100k
IX	Delmas 73	383k
IX	Croix des Bouquets	9k
VI	**Miragoane**	6k
V	**Verrettes**	49k
III	Santo Domingo	2,202k
III	**Guantanamo**	273k

bold cities appear on map (k = x1000)

Shaking Intensity MMI

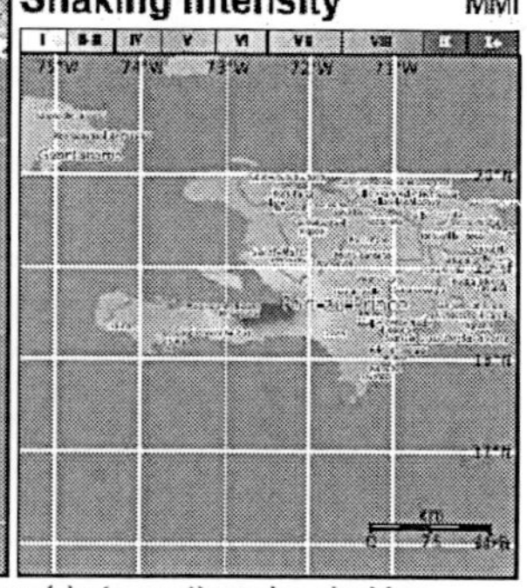

Overall, the population in this region resides in structures that are vulnerable to earthquake shaking, though some resistant structures exist. On June 24, 1984 (UTC), a magnitude 6.7 earthquake 329 km East of this one struck the Dominican Republic, with estimated population exposures of 320,000 at intensity VII and 2,964,000 at intensity VI, resulting in an estimated 5 fatalities. Recent earthquakes in this area have caused landslides that may have contributed to losses.

This information was automatically generated and has not been reviewed by a seismologist.

http://earthquake.usgs.gov/pager Event ID: us2010rja6

Source: USGS, http://earthquake.usgs.gov/earthquakes/pager/events/us/2010rja6/onepager.pdf.
Note: This is version 7 of the PAGER output, accessed on January 14, 2010.

Figure 5. Example of PAGER Output for the January 12, 2010, Magnitude 7.0 Haiti Earthquake.

RESEARCH—UNDERSTANDING EARTHQUAKES

U.S. Geological Survey

Under NEHRP, the USGS has responsibility for conducting targeted research into improving the basic scientific understanding of earthquake processes. The current earthquake research program at the USGS covers six broad categories:[50]

- *Borehole geophysics and rock mechanics*: studies to understand heat flow, stress, fluid pressure, and the mechanical behavior of fault-zone materials at seismogenic[51] depths to yield improved models of the earthquake cycle;
- *Crustal deformation*: studies of the distortion or deformation of the earth's surface near active faults as a result of the motion of tectonic plates;
- *Earthquake geology and paleoseismology*: studies of the history, effects, and mechanics of earthquakes;
- *Earthquake hazards*: studies of where, why, when, and how earthquakes occur;
- *Regional and whole-earth structure*: studies using seismic waves from earthquakes and man-made sources to determine the structure of the planet ranging from the local scale, to the whole crust, mantle, and even the earth's core; and
- *Strong-motion seismology, site response, and ground motion*: studies of largeamplitude ground motions and the response of engineered structures to those motions using accelerometers.

National Science Foundation

NSF supports fundamental research into understanding the earth's dynamic crust. Through its Earth Sciences Division (part of the Geosciences Directorate), NSF distributes research grants and coordinates programs investigating the crustal processes that lead to earthquakes around the globe.52 In 2003, NSF initiated a Major Research Equipment and Facilities Construction (MREFC) project called EarthScope that deploys instruments across the United States to study the structure and evolution of the North American Continent, and to investigate the physical processes that cause earthquakes and volcanic eruptions.53 EarthScope is a multi-year project begun in 2003 that is funded by NSF and conducted in partnership with the USGS and NASA.

EarthScope instruments are intended to form a framework for broad, integrated studies of the four-dimensional (three spatial dimensions, plus time) structure of North America. The project is divided into three main programs:

- *The San Andreas Fault Observatory at Depth (SAFOD)*, a deep borehole observatory drilled through the San Andreas fault zone close to the hypocenter of the 1966 Parkfield, CA, magnitude 6 earthquake;
- *The Plate Boundary Observatory (PBO)*, a system of GPS arrays and strainmeters[54] that measure the active boundary zone between the Pacific and North American tectonic plates in the western United States; and

- *USArray*, 400 transportable seismometers that will be deployed systematically across the United States on a uniform grid to provide a complete image of North America from continuous seismic measurements.

Through its Engineering Directorate, NSF funds the George E. Brown Jr. Network for Earthquake Engineering Simulation (NEES), a project intended to operate until 2014, aimed at understanding the effects of earthquakes on structures and materials.[55] To achieve the program's goal, the NEES facilities conduct experiments and computer simulations of how buildings, bridges, utilities, coastal regions, and materials behave during an earthquake.

CONCLUSION

At present earthquakes can be neither accurately predicted nor prevented, and in its 1990 reauthorization NEHRP shifted its program emphasis from prediction to hazard reduction. The program's focus has been on understanding the earthquake hazard and its risk to populations and infrastructure in the United States, developing effective measures to reduce earthquake hazards, and promoting the adoption of earthquake hazards reduction measures in vulnerable areas. In the 111th Congress, legislation introduced to reauthorize NEHRP, H.R. 3820, reemphasizes that approach but casts it in terms of hazard *mitigation* by stating that a major goal for the program should be "to reduce the loss of life and damage to communities and infrastructure through increasing the adoption of hazard mitigation measures." The bill further emphasizes the social aspects of mitigating earthquake hazards, calling for research to better understand institutional, social, behavioral, and economic factors that influence how risk mitigation is implemented, in addition to the traditional research into understanding how, why, and where earthquakes occur.

The emphasis on mitigation presented by H.R. 3820 reflects at least two fundamental challenges to increasing the nation's resiliency to earthquakes, and to most other major natural hazards such as hurricanes and major floods. The first is to assess whether social, behavioral, and economic factors can be understood in sufficient degree to devise strategies that influence behavior to mitigate risk posed by the hazard. Put simply, what motivates people and communities to adopt risk mitigation measures that address the potential hazard? A second challenge, which is more squarely an issue for Congress, is how to measure the effectiveness of NEHRP more quantitatively. It is inherently difficult to capture precisely, in terms of dollars saved or fatalities prevented, the effectiveness of mitigation measures taken before an earthquake occurs. A major earthquake in a populated urban area within the United States would cause damage, and a question becomes how much damage would be prevented by mitigation strategies underpinned by the NEHRP program.

A precise relationship between earthquake mitigation measures, NEHRP and other federal earthquake-related activities, and reduced losses from an actual earthquake may never be possible. However, as more accurate seismic hazard maps evolve, as understanding of the relationship between ground motion and building safety improves, and as new tools for issuing warnings and alerts such as ShakeMap and PAGER are devised, trends denoting the effectiveness of mitigation strategies and NEHRP activities may emerge more clearly. Without an ability to precisely predict earthquakes, Congress is likely to face an ongoing challenge in determining the most effective federal approach to increasing the nation's resilience to low-probability but highimpact natural hazards, such as major earthquakes.

End Notes

[1] ANSS is a nationwide network of seismographic stations operated by the USGS. GSN is a global network of stations coordinated by the Incorporated Institutions for Seismology. NEES is an NSF-funded project that consists of 15 experimental facilities and an IT infrastructure with a goal of mitigating earthquake damage by the use of improved materials, designs, construction techniques, and monitoring tools.

[2] U.S. House, Committee on Science, *National Earthquake Hazards Reduction Program Reauthorization Act of 2003*, H.Rept. 108-246 (Aug. 14, 2003), p. 13.

[3] Lifelines are essential utility and transportation systems.

[4] U.S. Congress, House Committee on Science and Technology, Subcommittee on Technology and Innovation, *Reauthorization of the National Earthquake Hazards Reduction Program: R&D for Resilient Communities*, testimony of Dr. Michael Lindell, 111th Cong., 1st sess., June 11, 2009.

[5] Ibid. Tornadoes could be considered similar to earthquakes as "rapid onset disasters" that provide little or no warning, but which elicit similar societal responses, according to Dr. Lindell.

[6] The current interagency coordinating committee for NEHRP submitted a strategic plan to Congress in October 2008. It is available at http://www.nehrp.gov/pdf/strategic_plan_2008.pdf.

[7] U.S. Geological Survey, Dept. of the Interior, *Earthquake Hazards—A National Threat*, Fact Sheet 2006-3016, March 2006, http://pubs.usgs.gov/fs/2006/3016/2006-3016.pdf. During the period 1975-1995, only four states did not experience detectable earthquakes: Florida, Iowa, North Dakota, and Wisconsin. See USGS Earthquake Hazards Program, *Earthquake Facts*, at http://earthquake.usgs.gov/learn/facts.php.

[8] *Seismometers* are instruments that measure and record the size and force of seismic waves, essentially sound waves radiated from the earthquake as it ruptures. Seismometers generally consist of a mass attached to a fixed base. During an earthquake, the base moves and the mass does not, and the relative motion is commonly transformed into an electrical voltage that is recorded. A *seismograph* usually refers to the *seismometer* and the recording device, but the two terms are often used interchangeably.

[9] See USGS "Earthquakes Facts and Statistics" at http://neic.usgs.gov/neis/eqlists/eqstats.html#table_2.

[10] Estimates of earthquake-related fatalities vary, and an exact tally of deaths and injuries is rare. For more information on the difficulties of counting earthquake-related deaths and injuries, see http://earthquake.usgs.gov/regional/world/ casualty_totals.php.

[11] U.S. Geological Survey, *Earthquakes with 1,000 or More Deaths Since 1900*, at http://earthquake.usgs.gov/earthquakes/world/world_deaths.php. This estimate does not include fatalities from the February 27, 2010, magnitude 8.8 Chilean earthquake, which has resulted in widespread destruction but few fatalities compared to the Indonesian, Pakistan, and Haiti earthquakes.

[12] Insurance Information Institute, http://www.iii.org/media/facts/statsbyissue/hurricanes/. Loss estimates are in 2005 dollars.

[13] Risk Management Solutions (RMS), Newark, CA, press release (Sept. 2, 2005), at http://www.rms.com/NewsPress/ PR_090205_HUKatrina.asp.

[14] FEMA Publication 366, *HAZUS MH Estimated Annualized Earthquake Losses for the United States* (April 2008), at http://www.fema.gov/library/ viewRecord.do?id=3265. Hereafter referred to as FEMA 366.

[15] A. M. Best Company Inc., *2006 Annual Earthquake Study: $100 Billion of Insured Loss in 40 Seconds* (Oldwick, NJ: A.M. Best Company, 2006), p. 12. The A. M. Best report includes estimates from catastrophe-modeling companies of predicted damage from hypothetical earthquakes in Los Angeles, the Midwest, the Northeast, and Japan. The report cites an estimate by one such company, Risk Management Solutions (RMS), that a hypothetical 7.4 magnitude event along the Newport-Inglewood Fault near Los Angeles would cause $549 billion in total property damage. A hypothetical 6.5 magnitude earthquake along a fault between Philadelphia and New York City would produce $901 billion in total loss, according to an RMS estimate.

[16] Building inventory refers to four main inventory groups: (1) general building stock, (2) essential and high potential loss facilities, (3) transportation systems, and (4) utility systems (FEMA 366).

[17] FEMA 366, p. 37.

[18] Ibid., p. 32.

[19] FEMA 366, p. 32 and p. 36.

[20] Ibid., p. 36.

[21] Ibid., p. 10.

[22] The largest earthquakes in New York, New Jersey, and Massachusetts were, respectively: 1944, Massena, NY, magnitude 5.8, felt from Canada south to Maryland; 1783, New Jersey, magnitude 5.3, felt from New Hampshire to Pennsylvania; and 1755, Cape Ann and Boston, MA, intensity of VIII on the Modified Mercalli Scale, felt from Nova Scotia to Chesapeake Bay (USGS Earthquake Hazards Program).

[23] USGS Circular 1188, Table 3.

[24] In contrast to California, where earthquakes occur on the active margin of the North American tectonic plate, the New Madrid seismic zone is not on a plate boundary but may be related to old faults in the interior of the plate, marking a zone of tectonic weakness.

[25] Andrew Newman et al., "Slow Deformation and Lower Seismic Hazard in the New Madrid Seismic Zone," *Science*, v. 284 (April 23, 1999), pp. 619-621.

[26] Seth Stein, Joseph Tomasello, and Andrew Newman, "Should Memphis Build for California's Earthquakes?" *Eos*, v. 84, no. 19, (May 13, 2003), pp. 177, 184-185.

[27] Seth Stein, "Code Red: Earthquake Imminent?" *Earth*, vol. 54, no. 1 (January 2009), pp. 52-59.

[28] The GSN also monitors nuclear explosions.

[29] USGS Earthquake Hazards Program, at http://earthquake.usgs.gov/research/monitoring/anss/.

[30] USGS FY2010 Budget Justification, at http://www.usgs.gov/budget/2010/greenbook/ FY2010_USGS_ Greenbook.pdf, p. I-10.

[31] Ibid., p. I-11.

[32] Email from William Leith, Advanced National Seismic System Coordinator, USGS, December 22, 2009.

[33] USGS FY2010 Budget Justification, pp. T-32 and T-33.

[34] USGS FY2010 Budget Justification, pp. T-32 and T-33.

[35] The number of stations necessary to generate a data-based ShakeMap depends on the urban area and geology, but roughly correspond to about half the number of planned stations per urban area, at a spacing of about 20 kilometers between stations. Personal communication, William Leith, Advanced National Seismic System Coordinator, USGS, January 11, 2010.

[36] Strong motion seismometers, or accelerometers, are special sensors that measure the acceleration of the ground during large (>6.0 magnitude) earthquakes.

[37] $9 million reflects the combined appropriations for USGS and NSF for FY2009. ARRA (P.L. 111-5) enabled approximately $10 million to be made available via the USGS and NSF for the GSN through FY2010. Email from William Leith, Advanced National Seismic System Coordinator, USGS, Dec. 21, 2009.

[38] IRIS is a university research consortium, primarily funded by NSF, that collects and distributes seismographic data.

[39] USGS Open-File Report 2004-1390, and California 24-hour Aftershock Forecast Map, at http://pasadena.wr.usgs.gov/step/.

[40] The California Integrated Seismic Network is the California region of ANSS; see http://www.cisn.org/.

[41] Earthquakes typically occur in clusters, in which the earthquake with the largest magnitude is called the main shock, events before the main shock are called foreshocks, and those after are called aftershocks. See also http://pasadena.wr.usgs.gov/step/aftershocks.html.

[42] The *epicenter* of an earthquake is the point on the earth's surface directly above the hypocenter. The *hypocenter* is the location beneath the earth's surface where the fault rupture begins.

[43] Stuart Simkin, NEIC, Golden, CO, telephone conversation, Nov. 4, 2006.

[44] In early 2006, the NEIC implemented an around-the-clock operation center and seismic event processing center in response to the Indonesian earthquake and resulting tsunami of December 2004. Funding to implement 24/7 operations was provided by P.L. 109-13.

[45] ShakeMaps for some areas outside the United States are also available. See http://earthquake.usgs.gov/eqcenter/ shakemap/.

[46] See the USGS Earthquakes Hazards Program for more information, at http://earthquake.usgs.gov/ earthquakes

[47] See also USGS Fact Sheet 2007-3101 at http://pubs.usgs.gov/fs/2007/3101/.

[48] See http://www.fema.gov/plan/prevent/hazus/hz_overview.shtm.

[49] See http://www.hazus.org/.

[50] See http://earthquake.usgs.gov/research/.

[51] Seismogenic means capable of generating earthquakes.

[52] See http://www.nsf.gov/div/index.jsp?div=EAR.

[53] See http://www.earthscope.org/.

[54] A strainmeter is a tool used by seismologists to measure the motion of one point relative to another.

[55] A non-profit NEES consortium (NEESinc.) has operated the facilities for the 10-year operating lifespan at the following institutions: Cornell University; Lehigh University; Oregon State University; Rensselaer Polytechnical Institute; University of Buffalo-State University of New York; University of California-Berkeley; University of California-Davis; University of California-Los Angeles; University of California-San Diego; University of California-Santa Barbara; University of Colorado-Boulder; University of Illinois at Urbana-Champaign; University of Minnesota; University of Nevada-Reno; and University of Texas at Austin. See http://www.nees.org/.

In: Haiti: Earthquake and Response
Editor: Rachael A. Donlon ISBN: 978-1-61668-998-8

Chapter 5

OPENING STATEMENT FOR HEARING ON HAITI, U.S. SENATE COMMITTEE ON FOREIGN RELATIONS, CHAIRMAN JOHN F. KERRY*

Washington, DC– Senate Foreign Relations Committee Chairman John Kerry (D-MA) today delivered the following opening statement at the hearing on Haiti:

Today, Haiti is reeling in the aftermath of what may well be the worst humanitarian catastrophe the Americas have ever seen. Well over 100,000 dead, and more dying every day. An estimated one million Haitians displaced. Large parts of Port-au-Prince and several outlying cities flattened. An already weakened infrastructure basically collapsed.

The numbers simply can't explain the horrors that millions of Haitians are living though. Instead, we begin to understand Haiti's tragedy through stories and images: a tent city next to a crumbled presidential palace; a Haitian child dividing one rationed meal among eight members of his family; a seventy-year old woman rescued from the cathedral in Port-au-Prince seven days after the earthquake—too weak to stand, but strong enough to sing church hymns as she was carried out on a stretcher.

It is impossible not to be moved by the suffering, as well as the resilience and dignity, of the Haitian people. It is our duty as neighbors and frankly as fellow human beings, to respond to their tragedy. And that responsibility doesn't end with rescue. We must help Haiti to rebuild in a way that leaves Haiti better off and better prepared the next time a natural disaster strikes.

Since the quake, America and the world have rushed in with as much assistance as Haiti's infrastructure permitted—quickly deploying search and rescue teams, food, water, medical equipment, shelter, and several thousand troops. We are also well aware of the suffering and heartbreak that has affected the hardworking Haitian-American community. Indeed, my home state of Massachusetts is home to the third-largest Haitian community in the United States, many of whom have lost relatives and close friends – we must do what we can to help.

There has been a tremendous outpouring of generosity from Americans and the international community. People have opened their wallets and their homes. I have been working closely with dozens of families in Massachusetts to expedite adoptions of Haitian

---

* This is an edited, reformatted and augmented version of the U.S. Senate Committee on Foreign Relations publication, dated January 28, 2010.

orphans that were under way before the earthquake. We are also working to make sure our government's relief efforts provide for the thousands of Haitian children orphaned or displaced by the earthquake, within the safeguards of the formal processes that protect the children from trafficking.

I would like to commend Dr. Shah, Secretary Clinton, and countless other Americans inside government and out who have made an impressive effort that all of us can be proud of. I would particularly like to honor U.S. diplomat Victoria DeLong, who lost her life, and to recognize the enormous loss suffered by the UN. The UN has made a massive contribution in Haiti over the years and, when the earthquake struck, they lost many outstanding people on the ground. We offer our profound gratitude and our condolences to the UN and to the families of the deceased. We also mourn the loss of Britney Gengel, and the agony her parents, Leonard and Cheryl Ann, have gone through trying to recover their daughter's body.

Of course the task before all of us is far from over.

First, we must continue the enormous ongoing effort to meet Haitians' immediate need for food, water, shelter, electricity, and emergency medical care. So far, thanks to UN peacekeepers and U.S. forces, the security situation has allowed these efforts to proceed in general calm.

Second, we need to use this humanitarian crisis to begin reversing the poverty and environmental degradation that plagued Haiti long before this tragedy. We cannot be satisfied to simply return Haiti to the unsustainable conditions of the past.

On January 11th, Haiti was already the poorest country in the Western Hemisphere. Even before the quake there were 380,000 orphans in Haiti. Most Haitians lived on less than a dollar a day; one in eight children died before their fifth birthday, and 40% were not enrolled in school; 120,000 Haitians were HIV-positive; and rural Haitians were plagued by malnutrition.

We need to help Haitians build a sustainable foundation – physical, social and economic – for a stronger and more stable society. This is a chance for Haitians to re-imagine their country as they rebuild it.

We must use every opportunity to help Haiti improve its living standards. Haiti has duty-free, quota-free access to the US market, a large pool of low-cost labor, and a large, hardworking North American Diaspora sending money home. Haiti was making steps toward recovery when the earthquake struck, and violent crime was declining. Haiti's progress will be more sustainable if its government takes a serious look at longer-term challenges such as environmental devastation and runaway population growth.

Third, Haiti's recovery must belong to the Haitian people. They may need our help today, but they must be empowered to build their own future down the road. President René Préval and Prime Minister Jean-Max Bellerive need to lead the national recovery, and civil society and democratic institutions must be protected and nurtured. Haiti's long-term success depends on a government that can inspire its people, work with the private sector, attract investment, and marshal resources to provide basic services, security, and rule of law.

Some have said that Haiti is a lost cause. Based on all I know of the Haitian people, and I've learned a lot more in recent days, I couldn't disagree more. Even in the darkest hours after the earthquake, Haitians who were poor to begin with—and then lost everything—reached out to help each other. They searched for missing neighbors and strangers, provided comfort and shelter and shared their meager food.

Looting and violence here and there may make headlines, but it is the Haitians' determination and decency in the face of disaster that will make the country's future. Schools may have collapsed, but Haitians' commitment to education will not.

Elisabeth Debrosse Préval, an economist and the president's wife, urged the Haitian people to ``stand up again and move forward." As they do, America will be there to help.

We are fortunate to have with us today three impressive witnesses with deep knowledge of Haiti and the challenge that we and the Haitian people will face.

In: Haiti: Earthquake and Response
Editor: Rachael A. Donlon
ISBN: 978-1-61668-998-8

Chapter 6

OPENING STATEMENT FOR HEARING ON HAITI, DICK LUGAR, U.S. SENATOR FOR INDIANA

U.S. Senate Foreign Relations Committee Ranking Member Dick Lugar made the following statement at today's hearing.

I thank the Chairman for calling this important hearing on the rescue, recovery, and longer term reconstruction efforts in Haiti. The unimaginable devastation caused by the earthquake and ongoing aftershocks since January 12, 2010, is one of the worst natural disasters to confront the Western Hemisphere in modern history. Nearly 3 million people have been affected, and authorities estimate that more than 150,000 people have died. My thoughts are with all those who have suffered loss.

The crisis situation in Haiti has the potential to destabilize security in the Caribbean. The social instability in Haiti represents a critical concern for the Dominican Republic and could have far-reaching implications if deteriorating conditions induce a mass exodus of Haitians by land and sea.

Immediate action by the United States to provide emergency assistance to Haiti was clearly warranted. The heroic efforts of American relief workers and the international teams of first responders are deeply inspiring. I commend the administration and many of my Senate colleagues who have worked to advance policies and legislation that will hasten Haiti's recovery.

Last week, I called on the Secretary of Homeland Security to grant 18 months of Temporary Protective Status for Haitian immigrants already residing in our country and to grant visa parole for orphans in the midst of adoption proceedings with American parents. The Secretary's quick action on both provisions will ensure that many vulnerable children are united with loving families and that all people of Haitian descent in the United States are in a position to contribute to Haiti's recovery.

A senior economist at the World Bank has projected that Temporary Protective Status could generate an additional $360 million in remittances sent to Haiti in 2010 – on top of the more than $1 billion transferred each year since 2006.

I am working with Senator Dodd on a bill that will encourage the IMF to provide debt relief to Haiti and ensure that IMF gold sale surpluses are used for low income countries, including Haiti. The legislation also would explore ways to invigorate economic activity in the country by adjusting U.S.-Haiti trade agreements.

Despite strong support from the United States, sustained international participation in Haiti is vital for its recovery. It is especially important that the international community provide governance assistance to the Haitian people.

The failures and corruption of past Haitian governments contributed greatly to the stress felt by the Haitian people before the earthquake, and the limitations of the current government constrain the prospects for recovery.

These harsh realities, compounded by the significant loss of life that has weakened the government and other institutions in Haiti, compel the international community to consider creative measures. Because of the devastation, Haiti's condition approximates that of failed state.

We should consider an enhanced role for the United Nations in the daily operations of the Haitian government until the country is stable and less dislocated. This would include the provision of food and shelter, reconstruction activities, budgetary affairs, security, and other aspects of governance vital to the Haitian people. The United Nations has the credibility and capacity to perform this role.

The relationship between the United Nations and the Haitian government should be a consensual, cooperative arrangement that preserves Haitian participation in decisions, while ensuring that the resources and expertise of the international community are brought to bear on the daily problems of Haiti. If the UN plays an increased security and recovery role, I believe Haitians will more quickly gather their bearings and begin to rebuild their lives and their country.

A top priority must be developing and implementing a plan to resettle temporarily the millions of internally displaced people, for whom Port-au-Prince has little to offer. Adequate food, water and shelter must be delivered in earnest to these resettlement areas.

I would also encourage the Obama administration to coordinate Haitian-American volunteers into Hometown Associations (HTAs) – small groups of volunteers in the United States who are linked with their former hometowns and institutions in Haiti.

Members of HTAs can serve as interpreters, support the temporary resettlement of refugees, and provide assistance to Haitians who are evacuated to the U.S. for medical treatment not available on the island.

The Haitian-American community is eager to be involved in the recovery of Haiti, but much greater coordination is required to put their skills to full use. Our government should explore ways to partner with NGOs and existing social networking platforms like Facebook and Twitter to facilitate this coordination. Organizing HTAs will increase our ability to leverage the flow of the more than $1 billion remitted from the United States.

I look forward to the insights of our distinguished panel of witnesses and innovative recommendations on strategies for moving Haiti forward.

In: Haiti: Earthquake and Response
Editor: Rachael A. Donlon
ISBN: 978-1-61668-998-8

Chapter 7

TESTIMONY OF DR. PAUL FARMER TO THE U.S. SENATE COMMITTEE ON FOREIGN RELATIONS 27 JANUARY 2010

Thank you for inviting me to testify today before the Senate Committee on Foreign Relations. I speak as the U.N. Deputy Special Envoy for Haiti —President Clinton, as you know, is the Special Envoy —and also as a physician and teacher from Harvard who has worked for over twenty-five years in rural Haiti. Today, my hope is to do justice to Haiti not by chronicling the events of the past two weeks, which are well known to you, but by attesting to the possibility of hope for the country, and of the importance of meaningful investment and sustainable development in Haiti.

That said, I will not pretend that hope is not at times difficult to muster.

As I was flying from Port-au-Prince to Montreal on Monday, headed to a conference on coordinating responses to the massive earthquake, I did the painful math in my hea d and counted close to fifty colleagues, friends, and family members who had lost their lives in the space of a minute.

The afternoon of the earthquake, several of my colleagues from Partners In Health and the UN, were, ironically, in Port-au-Prince for a meeting about disaster risk reduction. Partners In Health, through its Haitian sister organization, provides health care to the rural poor. By focusing on training and employing local talent, we have grown a great deal over the years. We are currently serving a population of well over 1.2 million and count about five thousand employees, most of them community health workers.

Of course, not all our colleagues survived. But the vast majority of them did survive, and they have spent the last two weeks working day and night to relieve the staggering suffering of the wounded and displaced. President Clinton, our colleagues, and I have been in the cities of Port-au-Prince, Jacmel,and Léogâne, as well as the less-affected Central Plateau and Artibonite Valley. Everywhere we have seen acts of great bravery and solidarity.

In addition to the heroism of friends and colleagues, I would like to note for the record the dignity and patience of the long-suffering Haitian people. During a visit last week to Haiti's largest teaching hospital, which shares a campus with the ruins of the nursing and medical schools, President Clinton remarked that no other people in the world would be so patient and calm in the face of so much suffering. This observation, though accurate, must not be misunderstood. People in Haiti are afraid not only for their options and futures, but still quite simply for their safety. A few nights ago, we sat in empty wards: hearing of impending aftershocks, the patients bolted outside with their IVs dangling from their arms. They refused, as have so many, to sleep inside the building—any building—but instead found tarpaulins and sheets, and lay down in the open courtyard.

This scene has repeated itself throughout the country and is a reminder of the logistics challenges facing all those who would be involved in the provision of shelter, clean water,and healthcare. The relief efforts, focused now on addressing the initial wave of devastation from the earthquake, will soon turn to a new set of concerns. Hastily cobbled-together camps are at risk of outbreaks of cholera and other waterborne disease. The Haitian government has wisely proposed avoiding huge camps, which will be difficult to manage, but we must hasten our efforts to get tents, tarpaulins, and latrines or composting toilets to Haiti. It is humbling to see the relief efforts be so slow—in large part because delivery of services was so weak before the quake. Now we must do more to get food and water to people every day for some time to come. Creating safe schools and safe hospitals, even makeshift ones, is a known need in rebuilding a society, and storm - resistant housing must also be a carefully considered priority since there is little time before the rainy season. Students need to be back in school;the planting season cannot be missed and requires fertilizer, seeds, and tools.

How will we fund such settlements, ongoing relief, the sowing of seeds, and the reconstruction that must follow? Major pledges have been made by the U.S., Canada, Japan, Spain, Brazil, the European Union, the Inter-American Development Bank, the World Bank, and others.[1] Indeed, *most* countries have responded. Even in far away and once-afflicted Rwanda, a group of community health workers making less than $200/month have been able to pull together $7000 in donations for their colleagues in Haiti. This is but a small portion of the billions needed, but hard to surpass as an eloquent testimony of human solidarity.

Even if adequate resources are available, the task before us will be extremely difficult.

Medical jargon, though at times arcane, can be helpful here. Today, Haiti is facing what we would term "acute on chronic" problems. Before January 12, the country was already facing huge long-term challenges in public health and education, the unemployment rate 2,3 over 70%, and a majority of its population was living on less than two dollars a day.

Food and water insecurity were already huge problems.Does this catastrophe create a chance for all of us to have a sounder, more solidarity-based relationship with Haiti? Or is it to be yet another chapter in a jeremiad of suffering and abuse of power ? In my last testimony here, in 2003, I expressed concern that the latter possibility was likely given our policies at that time. Today I will spend my time focusing on the potential for an entirely reconsidered relationship between the two oldest independent countries in the Americas: Haiti and my own.

Let me offer, as one example of the difficult relations between Haiti and the international community (and an echo of the nineteenth-century machinations I discussed in my last testimony before this committee), the donor conference I attended here in Washington last April. It was one of only two donor conferences I have ever attended, the second being in Montreal earlier this week. The results of the first are noteworthy and worrisome: despite $402 million pledged to support the Haitian government's Economic Recovery Program, when the country was trying to recover from a series of nat ural disasters resulting in a 15% reduction of GDP, it is estimated that a mere $61 million have been disbursed.[5] In the Office of the Special Envoy, we have been tracking the disbursement of pledges, and as of yesterday we estimate that 85% of the pledges made last year remain undisbursed.

Many of us worry that, if what's past is prologue, Haitians themselves will be blamed for this torpor. But as we have argued before, there are serious problems in the aid machinery, and these have contributed to the "delivery challenges " on the ground. The aid machinery currently at work in Haiti keeps too much overhead for its operations and still relies overmuch on NGOs or contractors who do not observe the ground rules we would need to follow to build Haiti back better. The fact that there are more NGOs per capita in Haiti than in any other country in the hemisphere is in part a reflection of need, butalso in part a reflection of overreliance on NGOs divorced from the public health and education sectors.[7]

Haiti will continue to need the contractors, and the NGOs and mission groups, but more importantly we will need to create new ground rules—including a focus on creating local jobs for Haitians, and on building the infrastructure that is crucial to creating sustainable economic growth and ultimately reducing Haiti's dependence on aid.

In other words, what we need is a way of "building back be tter" that strengthens governance but also strengthens the Haitian economy to provide for the needs of its people, especially the vast majority of Haitians who are desperately poor. There is an opportunity not only to build Haiti back better, but to build a more functional and beneficial aid structure. Over the past two decades, US aid policies have seesawed between embargoes and efforts to bypass governments, including elected ones not to Washington's taste.[8] Neither the international community nor the United States provided credible, long-term, financial investment in Haiti. Restructuring foreign aid and forgiving Haiti's crippling debts are essential to helping the country recover. US laws, including the Foreign Assistance Act of 1961 and its later revisions, prevent direct investment in the public sector; we will need to revisit these policies. Debt forgiveness is also needed to ease the financial drain that would otherwise hinder economic recovery and growth.

In building back Haiti, a credible body that has been working in Haiti could help to house a recovery fund. We need to commit funds and also to disburse them. To quote Jeff Sachs, "Haiti does not need a pledging session; it needs a bank account to fund its survival and reconstruction."[9]Such an account should be managed in collaboration with partners, the UN, and, of course, Haitian leadership, which would work directly and openly with partners to design and implement recovery plans coordinated at central and local levels. The effort must include a comprehensive post-disaster needs assessment, which should be supported by the US and other partners.

Might such plans work? In some of the darkest moments of the last two weeks, when the incapacity and lack of coordination of institutions on the ground was revealed repeatedly, I have thought often of Rwanda and what happened there in 1994. As a physician and teacher at Harvard, I have been lucky to work with the Clinton Foundation, Partners In Health, and the government of Rwanda on rebuilding health infrastructure in three of the last four districts that lacked central hospitals. As in rural Haiti, this has been a very positive experience. It has resulted in thousands of jobs for Rwandans, and has created a broadly accessible health care infrastructure—all with a modest price tag compared to traditional aid contractors.

If such progress can be made in Rwanda, which boasts strong leadership but in 1994 was the poorest country in the world,then one hopes it can be made elsewhere. In part because of this positive experience of working together with the Clinton Foundation in Rwanda (and in Malawi and Lesotho), I joined President Clinton six months ago as his deputy in the UN role he originated. As Special Envoy for Haiti, President Clinton has focused his attention not only on holding donors to the financial pledges they made, but also on reducing the risk of disasters and on job creation through the massive public works that are necessary to reforest Haiti, protect watersheds,and improve agricultural yield—the backbone of the Haitian economy. Private investment in Haitian businesses, especially small and medium-sized ones, is critical to rendering Haiti free of any dependence on aid—the goal of Rwanda by 2020, and moreover, a goal likely to be met.[12]Haiti also has, he noted, great potential as a tourist attraction but lacks the infrastructure to welcome the millions of people who should see Haiti's natural beauty and historic treasures like King Henri Christophe's Citadel.

This has been our mission: to build back better from the 2008 storms. We believe that these efforts were beginning to bear fruit. We had scheduled a meeting last week in the Hotel Montana to bring in another round of investors and also to discuss job creation. As many of you know, this hotel is no longer standing, and most inside it perished on January 12. But the need for such investments, and the need for public works that would create hundreds of thousands of jobs, remains.

If there is any silver lining to this cloud, it is that we can push job creation. It is a strange irony that supporters of economic assistance to Haiti are now obliged to shill for "cash for work " programs—for the quaint notion that people should be paid for their labor. Let us at least be honest: it is absurd to argue that volunt eerism and food-for-work programs will create sustainable jobs. But if we set the ground rules on reconstruction correctly, we will be able to create sustainable jobs.

In other words, if we focus the reconstruction efforts appropriately, we can achieve long - term benefits for Haiti. The UNDP is helping to organize programs of this kind, which should be supported and extended around the country. Putting Haitians back to work and offering them the dignity that comes with having a job and its basic protections is exactly what brought our country out of the Great Depression.

This was always the right thing to do, and aid programs persistently fail to get it right. So here is our chance: if even half of the pledges made in Montreal or other such meetings are linked tightly to local job creation, it is possible to imagine a Haiti building back better with fewer of the social tensions that inevitably arise as half a million homeless people are integrated into new communities.

Haiti needs and deserves a Marshall Plan—not the "containment" aspects of that policy, unless we are explicit about containing the ill effects of poverty, but the social-justice elements. But we need to be honest about the differences between post-war Europe and Haiti in 2010. Part of the problem, I've argued, is the way in which aid is delivered now as compared to in 1946—well before the term "beltway bandits"was coined. We need a reconstruction fund that is large, managed transparently, creates jobs for Haitians, and grows the Haitian economy. We need a reconstruction plan that uses a pro-poor, rights-based approach far different from the charity and failed development approaches that have marred interactions between Haiti and much of the rest of the world for the better part of two centuries.

Our country can be a big part of this effort. Debt relief is important, but only the beginning. Any group looking to do this work must share the goals of the Haitian people: social and economic rights, reflected, for example, in job creation, local business development, watershed protection (and alternatives to charcoal for cooking), access to quality health care, and gender equity. Considering all these goals together orients our strategic choices. For example, cash transfers to women, who hold the purse strings in Haiti and are arbiters of household spending, will have significant impact. This is a chance to learn and move forward and build on lessons learned in adversity—to build hurricane-resistant houses with good ventilation to improve air quality from stove smoke; to build communities around clean water sources; to reforest the terrain to protect from erosion and to nurture the fertility of the land for this predominantly agricu ltural country. It is the chance to create shelter, grow the local economy and incomes, and invest in improved health. This will do much to decrease the risk of another calamity, and to decrease the vulnerability of the poor—especially as we face the second wave of problems, including epidemic disease born of the earthquake.

As a doctor, I can tell you that bad infrastructure and thoughtless policy are visible in the bodies of the poor, just as are the benefits of good policy and well-designed infrastructure. In my almost 30 years in Haiti I have witnessed many political interventions and multiple coups. They have been unpleasant, even if their effects pale in the shadow of what we are now experiencing. Many people look at Haiti and despair. They say that aid is wasted, that there is no hope for this country. I would answer them with the positive experience of building Haitian-led programs in the Central Plateau and Artibonite Valley regions that have created five thousand jobs for people who would otherwise have no steady work. I advance this model not because it is associated with our efforts, but because job creation is the surest way to speed up the cash flow that is essential now. It is also the fastest way to make amends for our past actions towards Haiti, which have not always been honorable.

Sitting before you, I am at my core optimistic about the possibilities before us and the potential of our support to help rescue and transform our poorest neighbor. The response from citizens of the United States to the recent events in Haiti has been overwhelming and encouraging. There is the promise of solidarity by our leadership to make long-term commitments to the kinds of investments needed in Haiti—and to fulfilling them.

For two centuries, the Haitian people have struggled for basic human and economic rights, the right to health care, the right to education, the right to work, the right to dignity and independence. These goals, which Haitians share with people all over the world, should direct our polici es of aid and rebuilding.

As I wrote with colleagues in a recent op-ed—which is available in my written testimony—as physicians working in Haiti, we know first-hand that Haiti itself will soon be the casualty if we do not help build back better in the way envisioned by Haitians themselves.

Posted on Sat, Jan. 23, 2010
Tales from the front
BY DRS. PAUL FARMER, LOUISE IVERS AND CLAIRE PIERRE

The vocabulary of clinical medicine is large and arcane, but a couple of concepts are useful in diagnosing what is happening in Haiti and in setting a path. In the comi

acute-on-chronic"picture: Haiti's majority has long been dealing with serious problems and to this has been added the acute injury of a massive earthquake affecting much of the country, most notably its most heavily populated areas.

If any kind of chronology can be imposed on a disaster of this magnitude, we are moving into the next phase, where rescue and relief operations continue -- miraculous rescues of those trapped are still occurring, with one young girl and her brother pulled from rubble the other day and now recovering at the largest urban hospital --and are complemented by slowly coordinated efforts to bring food, drink, shelter, and basic medical services to the millions affected by the quake.

Some of the aid is starting to move, as repeat visits to Port-au-Prince's general hospital reveal: In the space of less than a week, the hospital, run by local staff, has been assisted by scores of surgical and medical volunteers and has moved from no functioning operating rooms to a dozen that are busy all day, every day and throughout the night, too.

This disaster has brought together goodwill and interest in Haiti such that for the first time in the country's history, there may soon be enough surgeons and trauma specialists.

There are, of course, many kinds of trauma, and even those who escaped unscathed physically have lost friends and loved ones, to say nothing of material possessions.

Across the country, as people continue to search for missing family members and friends, a kind of numbness is giving way to grief. Rescue workers and medical personnel and ad hoc logisticians, most of them Haitians, will need a break, as some of them have been working nonstop for over a week. One of our collaborators is still in the clothes in which she escaped with her life from her home.

Sense of Calm

Everywhere here you see Haitians helping each other. Despite reports of violence, what strikes many of us is the overall sense of calm: Former President Clinton, after bringing surgical supplies to the general hospital, noted that no other people in the world would be so patient and calm in the face of so much suffering.

A young Haitian colleague, already on the faculty of Harvard Medical School, is organizing scores of volunteers from every class. People have opened their homes and yards, which are covered with makeshift shelters: The chronic problem of housing in Haiti is now worsened by the acute problem of half a million newly homeless.

In addition to cross-class cooperation, it is clear that the Haitian diaspora, which scattered across North America and Europe (and even Rwanda, where a small group of Haitians is busy raising funds

-

at last this is

starting to look like a functioning hospital."

He and other Haitian professionals living in the United States -- Haitian physicians and nurses are a powerful force there -- have much to offer a large-scale rebuilding effort if it is coordinated with efforts to rebuild national institutions.

Another helpful notion from medicine do no harm."Knowing what not to do is not the same thing as knowing what to do -- who can be sure of what to do when nothing of this scale has been registered before? -- but it is important nonetheless to learn from years of international aid to Haiti.

First, long-term lack of coordination of relief and reconstruction efforts will be costly. Competition between self-described donor nations is worse than unhelpful. Even now, there is bedlam as medical teams arrive with excellent skills and intention, but insufficient coordination.

The many clinicians now in the country need to work together as a team.

One potential model of recovery for Haiti is the nation of Rwanda. After the 1994 genocide, Rwanda was overwhelmed by the international helping class, which included, in addition to many people of good will, a flock of trauma vultures, consultants and carpetbaggers. Under the strong leadership of the nascent government, including now-President Paul Kagame, leaders insisted that recovery and reconstruction aid be coordinated by the central and district governments. A number of nongovernmental organizations left Rwanda, but most would argue the decisions made then have helped to create a new model of collaboration between public and private actors, and contributed to Rwanda's remarkable post-genocide stability and growth.

The government of Rwanda has made a generous financial gift to the people of Haiti.

Second, neglecting the immediate-term needs in favor of the long view is a mistake. People need food, water, shelter and sanitation in the days and months to come, to complement the emergency medical care that has been dispatched.

Third, those who wish to help in the next few days would be wise to hold off on most in-kind donations. Some of these will surely be needed soon, but the best thing to do right now is to send cash to organizations that have deep connections to Haiti and can draw on local knowledge and local hands to respond to the immediate needs of the injured, homeless, and sick.

RESETTLEMENT EFFORTS

Fourth, we must do no harm in resettlement efforts. Housing will be an enormous challenge, and will require the best minds on the planet. We need to avoid creating intermediate -term camps that become slums.

Fifth, we must make sure that deportation of Haitians from the United States and elsewhere stops.

Prescriptions for Haiti will be bountiful from outside, but we must ensure that the prescriptions are correct. Haiti needs a different kind of assistance, one built on solidarity and respect and rooted in what the Haitian people want for themselves. Assistance offered now must develop food sovereignty for Haiti and investment in the rural area, now seeing an influx of those displaced from the capital.

The next few weeks will reveal some sense of the long-term prognosis for the reconstruction of Haiti. There is already talk of a $12 billion rebuilding tab.

Haiti needs and deserves a Marshall Plan. We need a reconstruction fund that is large, managed transparently, and creates jobs for Haitians, grows the Haitian economy and uses a rights-based approach that is pro-poor and based on something far different from the charity and failed development approaches that have marred interactions between Haiti and much of the rest of the world for the better part of two centuries.

As physicians working in Haiti, we know first-hand that Haiti itself will be the casualty soon if we do not help build back better in the way envisioned by Haitians themselves.

The authors are all physicians working with Partners In Health/Zanmi Lasante in Haiti and teach at Harvard Medical School.

Dr. Farmer serves as United Nations deputy special envoy for Haiti under Bill Clinton. www.haitispecialenvoy.org.

End Notes

[1] Walker, P. "Haiti earthquake aid pledged by country. "*Guardian.co.uk* 26 January 2010. Available at: http://www.guardian (Accessed January 27, 2010)

[2] Flintoff, Corey. "In Haiti, a low-wage job is better than none,"*All Things Considered,* June 14, 2009. Available at: http://www.npr.org/templates/story/story.php?storyId=104403034 (Accessed January 27, 2010)

[3] In 2006, the World Bank estimated that 78% of Haiti's 9 million people lived on less than $2 per day. See Haiti at a Glance , World Bank, Development Data Group (DECDG). Available at: http://siteresources.worldbank.org/INTHAITI/Resources/Haiti.AAG.pdf (Accessed January 26, 2010).

[4] For an overview of Haiti's water insecurity and past delays in international commitments to address this crisis, see Varma MK, Satterthwaite ML, Klasing AM et. al. Wòch nan soley: The denial of the right to water in Haiti. Robert F.Kennedy Memorial Center for Human Rights, Center for Human Rights and Global Justice at NYU School of Law, Partners In Health ,and Zanmi Lasante, 2008. Available at: http://www.chrgj.org/projects/docs/wochnansoley.pdf (accessed January 27, 2010).

[5] This estimate of disbursements was prepared in January 2010 in an internal memorandum of the UN Office of the Special Envoy For Haiti. Presi dent Clinton, in his capacity as UN Special Envoy, frequently appealed to donors to fulfill their commitments. See Helprin, J, "Bill Clinton chides nations over help to Haiti. "*Associated Press*,, September 9, 2009. Available at: http://www.newsvine.com/_news/2009/09/09/3243861-bill-clinton-chides-nations-over-help-for-haiti (accessed January 27, 2010).

[6] Farmer, P. "Challenging orthodoxies: The road ahead for health and human rights." *Health and Human Rights: An International Journal* 2008; 10(1): 5-19.

[7] Daniel, Trenton, "Bill Clinton tells diaspora: 'Haiti needs you now,'"*Miami Herald*, August 9, 2009. Available at: http://www.miamiherald.com/news/americas/haiti/story/1179067.html (accessed January 27, 2010).

[8] Farmer P, Smith Fawzi MC, and Nevil P. "Unjust embargo of aid for Haiti."*The Lancet* 2003; 361: 420 - 42 3 .

[9] Sachs, J. "After the earthquake, how to rebuild Haiti from scratch." *Washington Post,* 17 January 2010. Available at: http://www.washingtonpost.com/wp-dyn/content/article/2010/01/15/AR2010011502457.html (Accessed January 27, 2010)

[10] Republic of Rwanda, Ministry of Health, and Partners In Health. *African Rural Healthcare: An Evaluation of the South Kayonza, Rwanda Project (2005-2011).* Programme Report, Ministry of Health, 2006.

[11] United Nations Development Programme. *Human Development Report: 1997.* New York: Oxford University Press, 1997. 146-148. Available at: http://hdr.undp.org/en/media (accessed January 27, 2010).

[12] Republic of Rwanda. *Rwanda Vision 2020*. Kigali: Ministry of Finance and Economic Planning, Republic of Rwanda, 2002. Available at: http://www.cdf.gov.rw/documents%20library/important%20docs/Vision_2020.pdf (accessed January 27, 2010).

In: Haiti: Earthquake and Response
Editor: Rachael A. Donlon
ISBN: 978-1-61668-998-8

Chapter 8

RECONSTRUCTING HAITI, TESTIMONY OF JAMES DOBBINS, BEFORE THE SENATE FOREIGN RELATIONS COMMITTEE UNITED STATES SENATE JANUARY 28, 2010

In considering how best to help Haiti recover from the January 12 earthquake, it is important to recognize that one now has an international disaster relief operation superimposed on top of a preexisting post-conflict reconstruction mission. The earlier of these two operations began in 2004, when American and then United Nations troops assumed responsibility for security in Haiti.

Of the two operations, humanitarian relief is clearly the more urgent, but post conflict stabilization and reconstruction is ultimately the more important. One intends to restore Haiti, the other to transform it.

The ultimate aim of any post conflict mission is to leave behind a society better able to look after itself. Usually this means the ability to manage political domestic competition in ways that do not spill over into violent conflict. In Haiti's case, the objective must also be to improve that society's capacity to deal with the sort of natural disaster that, given this country's location, will continue to strike with some regularity.

In trying to help fix a failed or failing state, one must begin by analyzing the sources of fragility. The earthquake demonstrated the weakness of Haiti's infrastructure. It also highlighted the weakness of its governing institutions. This is the true source of Haiti's vulnerability to conflict and to natural disaster. In Haiti's case, state building, rather than nation-building is the more apt description of our mission.

The history of prior American and international interventions in Haiti must instill a sense of caution regarding the prospects for any transformation. Yet as a candidate for such assistance, Haiti has many advantages over other fragile states, including ones where the nation or state building process has yielded positive results. Most of those states were surrounded by conflict prone and predatory neighbors. Haiti sits in the midst of a zone of peace and relative prosperity. All of its neighbors are much richer, and none have any interest in destabilizing Haiti or inhibiting its development. Neither is Haiti divided by competing ethnic or religious groups. Haitians have a strong sense of national identity, and no serious sectarian divides. Haiti also has a large and relatively prosperous diaspora, many of whom are located at no great distance and enjoy frequent contact with their families on the island.

So Haiti does have certain inherent advantages. In addition, there are three newer factors which provide some hope that future efforts to help Haiti can yield more enduring results than those of the past.

First, the final departure of ex-President Aristide in 2004 has greatly diminished partisan rancor in both Port-au-Prince and Washington. At a moment when Bill Clinton and George W. Bush are campaigning together for relief to Haiti, one may hope that this American divide, which so hobbled American efforts to help Haiti in the past, has definitively closed.

Second, the outpouring of sympathy for Haiti as a result of the recent earthquake seems likely to yield a substantial increase in American and international aid levels. More money means more assistance and also more leverage to promote change.

Finally, the very immensity of the recent disaster has administered a shock to the Haitian political structure that can help ease resistance to reform and undermine longstanding barriers to progress.

My personal experience with Haiti dates back to the American intervention of 1994. This was one of five such nation-building operations with which I became associated, beginning with Somalia earlier in that decade, and ending with Afghanistan in the aftermath of 9/11. Since leaving government, we at RAND have issued a series of studies looking at the results of post conflict stabilization and reconstruction missions across each of these American led interventions, plus a larger number of UN led operations. Based on this body of work, I offer the following suggested guidelines for future aid to Haiti.

First, security is an essential prerequisite to reform, as it is to private investment. In the absence of security, any positive changes will eventually be washed away. Fortunately, Haiti is not a difficult society to secure. Contrary to the popular image, the Haitian population is neither heavily armed, nor inclined to violence. One has only to regard the patience with which the people of Port-au-Prince has awaited succor over the past two weeks to recognize its essentially peaceful nature. Haiti is no Somalia, Iraq, or Afghanistan. The comparative docility of the population is, in fact, one reason why very small numbers of armed men have on occasion been able to threaten the state and overturn governments. In 2004, for instance, Aristide was driven out of office by an insurgency numbering in the very low hundreds, equipped with nothing but small arms.

American troops are, therefore, unlikely to be required once the immediate humanitarian emergency passes. Securing Haiti should be well within the capability of the current UN peacekeeping force, modestly strengthened as it is being to help cope with the new, post earthquake challenges. Nevertheless, the United States can and should help the UN in this task by assigning an increased number of American police to the UN international police contingent. In doing so, the United States should draw, in particular, on Creole speaking Haitian American policemen from places like Miami and New York.

Second, stabilization and reconstruction operations take time. The 1994 American-led international intervention was a case in point. That operation was almost entirely successful in its own terms, but those terms were much too narrow. In launching the intervention, President Clinton promised to restore a freely elected President and then to keep American troops in Haiti only long enough to organize new elections, inaugurate a new President, parliament and local officials. He promised to do this all within the space of two years. This his administration preceded to do, hitting every benchmark, achieving every target, and suffering almost no casualties in the process. But two years was too short a time to fix a society as troubled as Haiti's. In the end the 1994 intervention accomplished little of lasting value.

Recent post-conflict stabilization and reconstruction operations have been lasting eight to ten years. The current effort in Haiti began in 2004, but the country has since suffered devastating flood damage and now the earthquake. The clock on this operation therefore needs to be set back to zero hour, and the UN Security Council should plan on keeping a peacekeeping force in Haiti for another eight to ten years.

Third, in a post-conflict environment, economic development and political reform programs need to be evaluated not just on their potential to promote growth and social justice, but for their capacity to ease tensions in the society and promote reconciliation between long hostile groups. In Haiti, these competing groups are not ethnic or religious but rather economic and social. Programs that might exacerbate such tensions should be scrapped or adjusted, in favor of those that draw competing groups into collaboration, even where the immediate economic payoff of such programs may be less. This means that programs to relieve poverty and create jobs will be a necessary part of any larger aid effort, even if their immediate impact is ephemeral, for without such visible signs of progress, significant elements of the population may be inclined to block longer term, larger pay off efforts at reform.

Fourth, assistance should be focused on building a more competent and efficient state. Haiti's vulnerability to natural disaster is not just a matter of weak building codes and poor infrastructure, but more fundamentally the result of having an exceptionally weak state that cannot provide even minimal public services – security, power, water, health and education – to the vast majority of its people. Haiti is, for instance, the only state in the entire Western Hemisphere that does not provide free primary education to most of its children.

The urgency of the immediate crisis requires that donor countries themselves provide people with food, water, medicine and shelter, bypassing the Haitian state. As we move beyond this emergency relief phase, the next priority will be to repair the country's most basic infrastructure – hospitals, schools, roads, electricity, telephones and government buildings. But these institutions should not be rebuilt on the old, inefficient, corrupt foundations. Rather the scale of this disaster offers the opportunity to accelerate long planned, oft delayed reforms in each of these sectors.

The port of Port-au-Prince has, for instance, long had the highest cost per ton in the hemisphere despite having the lowest wage rate. We should help rebuild this port, but not with the same grossly inefficient management and distorted cost structure. The same goes for the education ministry, the electric company, the telephone monopoly, the health ministry and the court system. Repair or replace the buildings, by all means, but also insist upon fundamental reforms in the management of these institutions.

Large amounts of American and other donor money will flow into Haiti in coming weeks. The temptation will be to spend most of it on American and foreign NGOs that can deliver essential services with fair reliability and good accountability, which Western legislators insist upon. But this sort of aid leaves behind no lasting local capacity to sustain those services. The second priority will be on bricks and mortar construction. This too will leave the underlying Haitian institutions unaltered. Such aid should, thus, be oriented to the extent possible on enhancing Haiti's capacity to govern. This means providing the Haitian government the wherewithal to hire well qualified staff at competitive wages, and programs to further train such staff and provide them the information systems and other support serviced needed to maximize their efficiency.

Fifth, the Haitian state should be built from the bottom up as well as the top down. This means assistance to mayors and local councils, and funding which will allow key government agencies to establish a presence beyond Port-au-Prince. For decades the population has moved off the land and into the big cities, particularly the capital. This exodus has now, as a result of the earthquake, been reversed. Assistance efforts should be designed to help those who have left the city to find a livelihood in the countryside, rather than return to the shanty towns from whence the have fled.

Sixth, the U.S. government needs to organize itself for a sustained high intensity effort to promote these reforms. The President and Secretary of State should invest a single individual with authority and responsibility for Haiti comparable to that Ambassador Richard Holbrook currently exercises for Afghanistan. Congress should authorize and appropriate new money for Haiti not in the usual categories of development assistance, security assistance, counternarcotics assistance, refugee assistance, etc., but in a single account unencumbered by earmarks and special limitations, and then work through the consultative and oversight processes with whomever the Administration designates as its point person to ensure that this money is carefully targeted and well spent.

Seventh, it is important that the international program to reform Haitian institutions not bear a made-in-Washington imprint. Large-scale international assistance will carry with it significant leverage to promote change, but this pressure needs to be exerted in a carefully calibrated manner. Candidate programs for reform need to be carefully chosen, local champions identified and empowered, local opponents co-opted, politicians lobbied and the public informed. The United Nations and the World Bank, the two major international institutions most heavily engaged in Haiti should be out in front, choosing and designing the necessary reforms and conditioning assistance on their achievement. The United Nations should continue to lead in reforming the security sector, to include police, courts and prisons, and in supporting elections and promoting political reform. The World Bank should assume leadership throughout the economic and social sphere, identifying the key changes needed and setting the conditions for assistance. The United States should work in concert with the other key donors, particularly Canada, France and the European Union, forming a small core group to quietly help the UN and the World Bank define their reform objectives and then working largely behind the scenes to ensure these objectives are achieved. The United States should contribute directly to UN and World Bank funds for Haiti, and should make sure that its own bilateral programs, and those of other donors contribute to, rather than undercut the reform programs set out by these institutions.

Finally, there are a couple of things that the United States is uniquely positioned to do by reason of its proximity to Haiti. These involve trade and immigration.

In 2006 Congress passed legislation providing Haiti uniquely generous but time limited access to the U.S. market. As with the UN peacekeeping mission, the time clock on this access should be set back to zero, recognizing that the earthquake has more than swept away whatever had been accomplished since these preferences originally went into effect.

The United States should also consider temporarily raising its annual quota for Haitian immigration. Haitian society may be economically dysfunctional, but Haitian immigrants have, quite to the contrary, proved to be hard working, family oriented, law abiding contributors to our society, even as they are one of the largest sources of support for those they leave behind in Haiti. Every dollar that they remit to relatives in Haiti is another dollar that does not need to come from the U.S. taxpayer. Expanding legal Haitian immigration thus seems a classic win-win proposition.

The current crisis, though tragic, offers the chance to boost Haiti out of decades of poverty and misrule. A successful strategy for doing so will require several elements: care in the design, sustained U.S. commitment, effective international coordination, and, above all, a focus on strengthening Haiti's governing institutions.

End Notes

[1] The opinions and conclusions expressed in this testimony are the author's alone and should not be interpreted as representing those of RAND or any of the sponsors of its research. This product is part of the RAND Corporation testimony series. RAND testimonies record testimony presented by RAND associates to federal, state, or local legislative committees; government-appointed commissions and panels; and private review and oversight bodies. The RAND Corporation is a nonprofit research organization providing objective analysis and effective solutions that address the challenges facing the public and private sectors around the world. RAND's publications do not necessarily reflect the opinions of its research clients and sponsors.

[2] This testimony is available for free download at http://www.rand.org/pubs/testimonies/CT339/.

In: Haiti: Earthquake and Response
Editor: Rachael A. Donlon
ISBN: 978-1-61668-998-8

Chapter 9

Haiti: From Rescue to Recovery and Reconstruction, M. Rony Francois, before the U.S. Senate Committee on Foreign Relations, United States Senate January 28, 2010

Good morning Chairman Kerry, Ranking Member Lugar and members of the Committee.

My name is Dr. Rony Francois. I am a physician and a doctor of Public Health by training. I am also the immediate past Secretary of Health for the State of Florida as well as the immediate past Director of Public Health for the State of Louisiana. Effective Monday, I will be the Director of Public Health for the State of Georgia.

I want to thank the Committee for inviting me to testify on the topic of "Haiti: From Rescue to Recovery and Reconstruction". My testimony will focus on the following: 1) Extent of the damages, 2) the immediate relief needs, 3) some intermediate goals, and 4) the long-term path to recovery and development.

Extent of the Damages:

An earthquake of 7.0 magnitude is a challenge for any country. But when it occurs in one of the world's most fragile countries, its intensity seems potentiated as it finds a destructive and sinister synergy in the vast landscape of feeble structures.

The estimated mortality is about 200,000. There are about 4,000 missing Americans. The number of injured approaches 1 million. The number of people affected is approximately 3 million. The economic damage is difficult to quantify.

My mother and father in law lost everything in Haiti. Fortunately, they were in the United States with me when the devastation occurred.

Immediate Relief Needs:

With the great number of people affected, the critical needs are food and water, medical care, temporary shelters, security and sanitation. In terms of diet, the most vulnerable groups are the newborns and young infants who require age-appropriate formula.

However, tensions in the Capital city and also the countryside continue to grow as many families do not have access to food and other basic goods. The escalation of frustration is already inducing violence and if resources are not more strategically distributed throughout the country, the security situation on the ground could rapidly deteriorate.

Medical teams must also remain vigilant about wound infections, sepsis, malaria, tetanus, respiratory infections, and typhoid as thousands more could die if conditions are not closely monitored and immediate steps are not taken to mitigate risks.

Intermediate Goals:

In light of over 50 aftershocks of a magnitude of 4.5 or greater, a thorough assessment of the structural integrity of remaining homes and buildings is absolutely critical for the future safety of the population. Before entering the U.S. and earning his doctorate degree in structural engineering - specializing in foundations and water resources - my brother worked as an engineer in Port-au-Prince and building codes at that time did not exist and construction practices were geared to withstand hurricane wind forces and not earthquakes.

In order to forecast the magnitude and hence the cost of debris/rubble removal and disposal, it is appropriate to compare this recovery effort to the models and lessons learned after the September 11, 2001 destruction at the World Trade Centers, after Hurricane Ivan's aftermath in Pensacola, Florida and after Hurricane Katrina's devastation of the Gulf Coast States – where recovery is still ongoing today.

The materials collected during this cleanup can be recycled or transported as fill for reclamation land along the coast of the Bay of Port-au-Prince.

Assuring proper drainage of rainwater will be a massive undertaking with the Mountain l'Hopital's steep watershed and the drastic reduction of pervious areas with tree cutting practices and anarchistic construction. The existing system was already obsolete and undersized and did not seem to rely on accurate hydrological data. In addition, with the earthquake's underground stresses generated from the wave propagation, it must be anticipated that the majority of structures and pipes (mostly concrete) have sustained irreversible damage.

The waterworks and sewer infrastructures gauged against current Environmental Engineering standards were already inadequate. Construction from scratch might be considered.

The source of the water supply might be adjusted after evaluation of the Cul-de-Sac Plain aquifer capacity and inventory of other reserves through groundwater hydrology. Another very viable option is to tap in the Bay of Port-au-Prince readily available sea saltwater through a desalination process.

Long-Term Reconstruction and Development:

Decentralization away from Port-au-Prince must be a primary goal of future development efforts. The earthquake has already triggered an exodus of the population to various parts of Haiti. In order for that migration to become permanent, investments must be made in agriculture, food security, local jobs, tourism infrastructure, security, roads, and airports.

Access to rudimentary medical care must also be decentralized with a network of community health centers across the country. Public health must become the foundation of this new health system and foster the widespread training and use of health promoters or agents of health.

Improvements in literacy should also remain central to any development efforts.

I would like to share with the Committee a few of the recommendations of the Haiti Advisory Group created by Executive Order on July 14, 2004 by former Florida Governor, Jeb Bush. The Advisory Group consisted of 17 prominent Haitian-American members with significant experience in Haitian issues. The Advisory Group formulated a set of 25 recommendations to the Governor and the State of Florida. These prophetic recommendations, which are still very relevant today, included a wide range of problematic issues that fall into the following categories:

- Security Recommendations
- Economic Development Recommendations
- Disaster Preparedness/Environmental Rehabilitation

Although Florida was the main partner for these projects, I would propose that these recommendations be adopted by the United Nations. Examples include:

- Creation of a dedicated Professional Exchange and Training Program with the Government of Haiti. This would allow volunteers, utilizing their vast experience and skills, to travel to Haiti to provide in-country technical assistance/training. The objective of the professional exchange and training program would be to provide Haitian participants with the knowledge, skills, and resources they need to work more effectively in their respective fields and to build greater capacity within Haitian institutions.

For example, Haiti produces only about 80 physicians each year, but that number will steeply decline since many of the medical schools were destroyed. What will happen to medical-students already enrolled? How will they complete their studies? Will opportunities to study abroad be made available? A sharp increase of injured and a sharp decrease in doctors creates a formula for an intractable disaster.

- The international community should utilize its great expertise and experience in law enforcement, corrections, and public safety to assist the growth and professionalization of the Haitian National Police (HNP).
- The United Nations could sponsor a Haiti Trade Mission and Matchmaker Fair in which representatives of Haitian businesses can travel abroad to meet with counterparts. Promotion of investment in Haiti, as well as networking and identification of export opportunities for Haitian products would be primary goals.
- The United Nations should begin now to assist Haiti's Ministry of Commerce, Industry, & Tourism in its plan to create a targeted tourism marketing campaign to attract Haitian-Americans and other visitors to their country as tourists in order to rediscover its natural beauty and historical attractions.
- Haiti's infrastructure is now exponentially more vulnerable to hurricane induced disasters like mud-slides and flooding. Hurricane season begins June 1st and immediate action must be taken to mitigate potential risks. The United Nations should employ its well-developed expertise and experience with natural disasters to provide technical assistance to the Republic of Haiti's Directorate of Civil Protection,

in order for this agency to develop appropriate disaster management infrastructure and training.

In closing, I extend fervent prayers to all of those who are affected by the earthquake as well as the first responders who have come from near and far. I also want to acknowledge the magnanimous generosity of those who have contributed to the relief efforts in Haiti. As we move forward, the United States and the international community must assure that investments made in rebuilding Haiti are actually carried out through community-based organizations, faith-based entities, non-governmental agencies, and non-profit organizations. My personal hope is that a new generation of leaders will emerge from these ashes to selflessly lead Haiti onto a new path of prosperity through integrity, hard work, transparency, perseverance and true democracy.

Thank You.

In: Haiti: Earthquake and Response
Editor: Rachael A. Donlon
ISBN: 978-1-61668-998-8

Chapter 10

TESTIMONY, "RECONSTRUCTING TO REBALANCE HAITI AFTER THE EARTHQUAKE", ROBERT MAGUIRE, PH.D. TRINITY WASHINGTON UNIVERSITY AND THE UNITED STATES INSTITUTE OF PEACE (USIP), TESTIMONY PRESENTED BEFORE THE SUBCOMMITTEE ON INTERNATIONAL DEVELOPMENT, FOREIGN ASSISTANCE, ECONOMIC AFFAIRS AND INTERNATIONAL ENVIRONMENTAL PROTECTION OF THE US SENATE COMMITTEE ON FOREIGN RELATIONS, FEBRUARY 4, 2010

Thank you for inviting me to testify today.

My first visit to Haiti was in 1974. My first full day in Haiti was the day that Haiti's National Soccer Team scored *the* incredible goal against Italy in the World Cup. That may not mean much to Americans, but to Haitians it means *everything*. I have come to take this coincidence as a sign that there was bound to be some kind of unbreakable bond between Haiti and me. And that came to pass.

My most recent visit ended on January 10, 2010, two days before the earthquake. In between, I have visited Haiti more than 100 times, as a US government official working with the Inter-American Foundation and the Department of State; as a scholar and researcher, and as a friend of Haiti and its people. I have traveled throughout that beautiful, if benighted, land. I have met and broken bread with Haitians of all walks of life. I have stayed at the now-destroyed Montana Hotel. I had dinner there 5 days before the quake, chatting with waiters and barmen I had befriended over the years. I speak Creole. I have lost friends and colleagues in the tragedy. I am anxious to share my views and ideas with you.

In the deep darkness of the cloud cast over Haiti by the terrible tragedy of January 12 there is an opportunity for the country and its people to score another incredible goal, not so much by reconstructing or rebuilding, but by restoring a balance to achieve a nation with less poverty and inequity, improved social and economic inclusion, greater human dignity, a rehabilitated environment, stronger public institutions, and a national infrastructure for economic growth and investment. And, if that goal is to be scored, relationships between Haitians and outsiders also will have to be rebalanced toward partnership and respect of the value and aspirations of all Haiti's people.

A Country Out-of-Balance

In the five decades that I have traveled to Haiti, I have seen the country become terribly out-of-balance. Much of this revolves around the unnatural growth of Haiti's cities, especially in what Haitians call "the Republic of Port-au-Prince." In the late 1970s, Haiti's rural to urban demographic ratio was 80% to 20%. Today it is 55% to 45%. The earlier ratio reflected what had been chiefly an agrarian society since independence. The population of Port-au-Prince in the late 1970's was a little over 500,000 – already too many people to be adequately supported by the city's physical infrastructure. By then, Haitians from the countryside had already begun trickling into the capital city as a result Dictator Francois "Papa Doc" Duvalier's (1957 – 1972) quest to centralize his grip on power. Under Papa Doc, a ferocious neglect beyond PAP took place, as ports in secondary cities languished, asphalted roads disintegrated and, in some cases, were actually ripped-up, and swatches of the countryside were systematically deforested under the guise of national security or by way of timber extraction monopolies granted to Duvalier's cronies. Small farmers were ignored as state-supported agronomists sought office jobs in the capital.

The only state institutions present in the countryside were army and paramilitary (*Tonton Makout*) posts and tax offices – which enforced what rural dwellers told me was a 'squeeze – suck' (*pese - souse*) system of state predation. With wealth, work and what passed for an education and health infrastructure increasingly concentrated in Port-au-Prince (PAP), it was no wonder that poor rural Haitians had begun to trickle off the land into coastal slums with names like "Boston" and "Cite Simon" (named after Papa Doc's wife), and onto unoccupied hillsides and ravines within and surrounding the city.

The trickle turned into a flood in the early 1980's when the rapacious regime of Jean-Claude "Baby Doc" Duvalier (1972 - 1986) yielded to Haiti's international 'partners' – governments, international financial institutions, and private investors – who had set their sights on transforming Haiti into the "Taiwan of the Caribbean." Political stability under the dictatorship combined with ample cheap labor and location near the US formed a triumvirate that shored up this idea, and the "'Tawanization" of Haiti proceeded, creating by the mid-1980's somewhere between 60K and 100K jobs in assembly factories, all located within PAP. Fueled by a parallel neglect of Haiti's rural economy and people, the prospect of a job in a factory altered the off-the-land trickle into a flood, as desperate families crowded the capital in search of work and the amenities – education, especially – that the city offered. Between 1982 and 2008, Port-au-Prince grew from 763,000 to between 2.5 and 3 million, with an estimated 75,000 newcomers flooding the city each year.[1]

As immigrants piled up in slums, on deforested and unstable hillsides, and in urban ravines, the opportunity offered by the city became a mirage. Following the ouster of Duvalier in 1986 when, as Haitians say, "the muzzle had fallen" (*babouket la tonbe*) and freedom of speech and assembly returned, factory jobs began to dry up as nervous investors sought quieter, more stable locations. By the 1990's, only a fraction of those jobs remained. Yet the poor continued to flow into the city.

Papa Doc's centralization, combined, under the rule of Baby Doc, with the urban-centric/Taiwanization policies of key donor countries (including the US) and international banks had a devastating impact on Haiti. Enacted by a government and business elite who saw these policies as a golden opportunity to make money, the internationally-driven Tawanization of Haiti neglected what Francis Fukuyama has pointed out as the key to Taiwan's own success: a necessary investment in universal education and agrarian improvement *before* investing in factories.[2] Haiti's people were viewed internationally and by local elites strictly as pliant and ample cheap labor. Education might make them ornery. Avoid it. Why invest in agriculture when cheaper food – heavily subsidized imported flour and rice – could feed Haiti's growing urban masses? In the late 1970's Haiti did not need to import food. Today, it imports some 55% of its foodstuff, including 360,000 metric tons of rice annually from the US.[3]

The folly of these policies was seen in early 2008, during the global crisis of rapid and uncontrolled commodity price increase, when that rice, still readily available, was no longer cheap and the urban poor took to eating mud cookies to survive. Another spinoff of these fallacious policies of the 80's was political instability. Poor Haitians took to the streets in early 2008 to protest "*lavi chè*" (the high cost of living), with the result being the ouster of the government headed by Prime Minister Jacques Edouard Alexis who, coincidentally, had just won praise from the US and the International Financial Institutions (IFIs) for the creation of a national poverty reduction and economic growth strategy that would serve as a blueprint for developing *all* of Haiti, and reversing the long history of rural neglect.

Rural neglect combined with migration to cities, moreover, placed considerable pressure on those still in the countryside to provide the wood and charcoal the burgeoning urban population required. Here was a recipe for desperately poor people to further ravish the environment. Today, 25 of Haiti's 30 watersheds are practically devoid of vegetative cover.

Port-au-Prince, and other cities, particularly Gonaives, had become disasters waiting to happen as a result of these developments. The vast majority of the 200,000 who perished on January 12th were poor people crowded on marginal land and into substandard housing devastated by the quake. The vast majority of the thousands who died in the floods in Gonaives in 2004 and 2008 were poor people crowded on alluvial coastal mud flats and in river flood plains. For Haiti's poor, their country has become a dangerous place and a dead-end. Is it surprising that Haitians seek any opportunity to look for life (*chèche layi*) elsewhere? As one peasant told me in the 1990's, 'we have only two choices: die slow or die fast. That's why we take the chance of taking the boats (to go to Miami)' (*pwan kantè*).

Haiti had lost its balance in other ways, particularly in social and economic equity, and in the ability of the state to care for its citizens. By 2007, 78 percent of all Haitians – urban and rural – survived on $2.00 a day or less, while 68 percent of the total national income went to the wealthiest 20 percent of the population.[4] During the 29 year Duvalier dictatorship, Haitian state institutions virtually collapsed under the weight of bad governance. Following the 1986 ouster of Baby Doc, who, ironically, was lorded with foreign funds that went principally to

Swiss bank accounts, donors were loathe to work with successor governments – including those democratically elected. Instead, they chose to funnel hundreds of millions annually through foreign-based NGOs that enacted 'projects' drawn up in Washington, New York, Ottawa, etc. and that lasted only as long as the money did. Haitians, by the early 1990s, were derisively calling their country a "Republic of NGOs" and by 2008, none other than the President of the World Bank, Robert Zoellick, lamented the cacophony of "feel good, flag-draped projects" that had proven a vastly inadequate substitute for a coherent national development strategy.[5]

Without doubt, Haiti was seriously out-of-balance before the earthquake and Port-au-Prince was a disaster waiting to happen. Many had feared that it would come by way of a hurricane; rather the earth shook. Now, let us see how we might make a positive contribution in restoring balance to Haiti so that when, inevitably, the country is struck by another natural disaster – be it seismic or meteorological – it is less vulnerable and better able to confront and cope with the disaster.

REBALANCING TO 'BUILD BACK BETTER'

Allow me to stress two points: we must be fully cognizant of past mistakes, such as those outlined above; and the key to 'building Haiti back better' is to work toward a more balanced nation with less poverty and inequities, less social and economic exclusion, greater human dignity, and a commitment of Haitians and non-Haitians toward these essential humanistic goals. With this, Haiti can also achieve and sustain a rehabilitated natural environment, stronger public institutions, a national infrastructure for growth and investment, and relationships between Haitians and outsiders that are based on partnership, mutual respect, and respect of the value and aspirations of all Haiti's people.

What follows are ideas and recommendations based on not only my experience in Haiti, but on endless discussions/conversations with Haitian interlocutors. In this regard, I should add that the principal reason for my visit to Port-au-Prince in early January was to deliver an address on prospects for rebalancing Haiti. That presentation was made to an audience of 50 or so Haitian civil servants and policy analysts – some of whom I fear are no longer with us - who gathered in Port-au-Prince at a Haitian think tank. My ideas were received by them with great and at times animated interest.

1. Welcoming Dislocated Persons: A de facto Decentralization

Since the quake, some 250,000 Port-au-Prince residents have fled the city, returning to towns and villages from which they had migrated or where they have family. An estimated 55,000 have shown up in Hinche in Haiti's Central Plateau; the population of Petite Riviere de l'Artibonite has swelled from 37,000 to 62,000; St. Marc's 60,000, has swollen to 100,000.[6] The flight of Haitians away from a city that now represents death, destruction and loss might become a silver lining in today's very dark cloud. If that is to be the case, however, we – both the government of Haiti and its international partners – must catch up with and get ahead of this movement. Already underdeveloped rural infrastructures and the resources of already impoverished rural families are being stretched. The provision of basic services to these displaced populations is an urgent priority. If conditions in the countryside are not improved, the displaced will ultimately return to Port-au-Prince, to replicate the dangerous dynamics of earlier decades.

To catch up and get ahead of this reverse migration, we should support an idea proffered by the Government of Haiti in Montreal last week: the reinforcement of 200 decentralized communities. As soon as possible, "Welcome Centers" might be stood up in towns and villages. They can be temporary, to be made permanent later. They can serve as decentralized 'growth poles' that offer multiple services, including relief in the short term, with health and education facilities attached. Let us not forget that Haiti has lost many of its schools and among those fleeing the devastated city are tens of thousands of students. Twenty five percent of Haitian rural districts do not have schools. And schools that exist outside Port-au-Prince are usually seriously deficient. The reverse migration we are seeing today offers a golden opportunity to rebalance the education and health system of Haiti.

The centers can coordinate investment and employment opportunities, as well as state services including robust agronomic assistance to farmers. Haiti's planting season is

almost here and now more than ever the country needs a bountiful harvest. Displaced people working as paid labor can reinforce Haiti's farmers. Infrastructure needs to be rebuilt – or built for the first time - including schools, health clinics, community centers, roads, bridges and drainage canals. Hillsides need rehabilitation, particularly with vegetative cover and perhaps even stone terraces. Providing work for not just the displaced, but to those they are joining in towns and villages throughout Haiti, will go a long way toward rebalancing Haitian economy and society, and toward repairing a social fabric ripped to shreds by decades of neglect and subsequent migration. This is an opportunity that must be seized.

2. Support the Creation of a National Civic Service Corps

Since 2007, various Haitian government officials and others have been working quietly on the prospect of creating a Haitian National Civic Service Corps. Citizen civic service is mandated in Article 52-3 of the Haitian constitution and, even before the quake, the idea of a civic service corps to mobilize unemployed and disaffected youth seemed attractive. Now is the time for this idea to take off. As I have recently written, a 700,000-strong national civic service corps will rapidly harness untapped labor in both rural and urban settings, especially among Haiti's large youthful population, to rebuild Haiti's public infrastructure required for economic growth and environmental rehabilitation and protection; increase productivity, particularly of farm products; restore dignity and pride through meaningful work; and give Haitian men and women a stake in their country's future. It will also form the basis of a natural disaster response mechanism.[7]

If this all sounds familiar, it should: the idea of a Haitian National Civic Service Corps parallels the same thinking that went into the creation of such New Deal programs as the Works Progress Administration (WPA) and the Civilian Conservation Corps (CCC).

We have seen what these programs did to help the United States and its people stand up during a difficult time.

In the aftermath of the storms that devastated Haiti in 2008, Haitian President Prèval asked not for charity, but for a helping hand to allow Haitians to rebuild their country. Today he is making a similar point. Here is more symmetry between the Haiti and the U.S. As Harry Hopkins, the legendary administrator of the Works Progress Administration, pointed out: "most people would rather work than take handouts. A paycheck from work didn't feel like charity, with the shame that it conferred. It was better if the work actually built something. Then workers could retain their old skills or develop new ones, and add improvements to the public infrastructure like roads and parks and playgrounds."[8] Let's help Haiti restore its balance by supporting a national civic service corps that can accomplish the same for Haiti and its people as our New Deal programs did in the United States decades ago.

To reiterate, as was the case with our New Deal, Haiti's civic service corps must be a 'cash-for-work' initiative. Cash-for-work will inject serious liquidity into the Haitian economy and stimulate recovery from the bottom-up. Already there are various entities employing Haitians in a variety of cash-for-work programs. This Monday, for example, the UNDP announced that it has enrolled 32,000 in a cash-for-work rubble removing program; a number expected to double by tomorrow.[9] Coordination of existing efforts within an envisaged national program will be essential to maximizing how Haiti can be built back better – by its own people, with everyone wearing the same uniform.

A special commission, similar to those established by President Prèval in 2007 to engage Haitians from diverse sectors to study and make recommendations on key issues confronting his government, might be established to oversee this coordination. (Other special commissions could be mounted to tackle other topics or needs and as a means of expanding the Haitian government's human resource circle.) Such a commission could be enlarged to include representatives of key donors. A central figure like Harry Hopkins will have to lead the endeavor. Perhaps such a figure could emerge from Haiti's vaunted private sector. In any case, let's avoid a repetition of the cacophony of feel good, flag-draped projects.

3. Strengthen Haitian State Institutions through Accompaniment, Cooperation and Partnership

At the Senate Foreign Affairs Committee hearing on Haiti held last week, witnesses spoke of the need to rebuild the Haitian state from the bottom up, and of working with Haitian officials – not pushing them aside. I agree with these points whole-heartedly. This is not the time to impose governance on Haiti – that is a 19th century idea unfit for the 21st century. This is an opportunity to help strengthen Haiti's public institutions, not to replace them.

As pointed out above, the capacity of the Haitian state, never strong to begin with, has deteriorated progressively over the past 50 years. In recent years that was due in part to international policies that circumvented state institutions in favor of private ones – both within Haiti and from beyond, and left the resource-strapped government virtually absent in the lives of its citizens. In the aftermath of the quake, we see starkly the results of the decimation of the Haiti state. The already weak state has been further set back by the death of civil servants and the loss of state facilities and physical resources. In this context, the government of President Rene Prèval and Prime Minister Jean Max Bellerive has taken much criticism for its response – or lack thereof – in the past few weeks.

It is easy to kick someone in the teeth when he or she is already on the mat. Rather than swinging our foot, however, we should offer our hand. This is the time of the Haitian government's greatest need. Achieving cooperation and partnership, as pointed out by

Canadian Prime Minister Harper at the recently-held Montreal Conference, is the biggest concern.[10] Over the past four years, the Prèval government has won praise internationally – and among most in Haiti – over its improved management of the affairs of the state. Political conflict, though still extant, has diminished considerably. Haiti's terribly polarized society is a little less polarized today. Moderation and greater inclusion – not demagoguery and a winner-takes-all attitude – have worked their way into the ethos of the Haitian political culture. Partnership to strengthen the Haitian state was on the horizon following the 'new paradigm for partnership' agreed to at the April 2009 Donors Conference.[11] Let's stay that course. Generations of bad governance and a zero sum political culture are not turned around overnight.

Quietly, but steadily in the post-quake period, the Haitian government has been picking itself up by its bootstraps beyond the photo-ops and glare of the cameras to reassemble, and then to reassert, itself.[12] Still, given the magnitude of this catastrophe, the government is overmatched. Any government would be. This is not the time to cast aspersions. It is the time to work in partnership and to accompany Haitian leaders through their time of loss and sorrow, into a more balanced and better future.

4. Get Money into the Hands of Poor People

In 1999, Peruvian economist Hernando de Soto estimated that there was $5.2 billion in 'dead capital' in Haiti, shared among 82 percent of the population. Of this sum, $3.2 billion was located in rural Haiti. This amount dwarfed by four times the total assets of Haiti's 123 largest formal enterprises. [13] This capital, principally in the hands of poor people in the form of property, land, and goods, is considered 'dead' because it cannot be used to leverage further capital for investment and growth. To free it up, clear titling would be required along with a reduction of red tape and corruption, and a brand new attitude toward Haiti's most vibrant form of capitalism – its' informal economy – and the poor entrepreneurs who make it work. Doubtless, you have seen post-quake stories of how Haiti's grassroots entrepreneurs began rebounding within days.

A key to Haiti's recovery – and, yes, to its rebalancing – is to get capital into the hands of grassroots entrepreneurs – be they still in Port-au-Prince or elsewhere in the country. Formalizing dead capital – which will be a long, tedious and conflictive path, but one that perhaps can be facilitated now through such steps as the issuance of provisional land and property titles that subsequently are fully formalized – is but one way of getting liquid assets into poor people's hands. Others, more expeditious, include:

- More small loans (microcredit) to entrepreneurs, particularly those who produce something, including farmers. Farmers with capital will not just produce more food, but will increase employment. Government studies indicate that a 10% increase in man-hours on farms will create 40,000 new jobs.[14] One strong candidate to improve microcredit throughout Haiti is an organization called FONKOZE. With more than 33 branches country-wide, it serves some 175,000 members, mostly among those who make – or made prior to their engagement with microcredit - $2.00 a day or less. FONKOZE also facilitates the efficient and lower cost decentralizing of the flow of funds sent to Haiti from family abroad.

- Haiti must now benefit from a conditional cash transfer (CCT) program. Brilliantly popular in such places as Mexico and Brazil, CCT programs serve as a means of transferring cash to the poorest of the poor, conditioned upon the children of poor families attending quality schools and fully operational clinics. Mexico's program is largely rural; Brazil's more urban-oriented. In both cases, they have succeeded in assisting millions of poor families improve living standards while sending their children to schools and clinics. As such, CCTs have invested in future human resources. Such a program in Haiti could accomplish these goals, but only if Haiti's educational and health systems are extended into rural areas (helping to rebalance) and upgraded in existing locations (helping to rebuild). Importantly, CCT programs provide the government with the challenge/opportunity of being a positive presence in the lives of citizens. In Haiti, this is essential as a means of enabling the government to move from being largely absent to being positively present in the lives of citizens, and to demonstrate therefore that there are tangible fruits of democratic governance.

5. Seek-out and Support Institutions, Businesses, and Leaders who work toward Greater Inclusion, Less Inequality, and Enact Socially-Responsible Strategies for Investing in Haiti

One cannot discuss the future of Haiti without considering the prospect of external investment to create factory jobs, particularly in view of the HOPE II legislation and its potential benefits. Beyond any doubt, factory jobs should be a part of Haiti's future. Already, some of the assembly plants in Port-au-Prince are back in operation, to the satisfaction of both owners and workers.[15] In this regard, support should be given to the "Renewing Hope for Haitian Trade and Investment Act for 2010" introduced by Senators Wyden and Nelson. But, as this legislation is considered, three important points must be kept in mind *if* this job creation strategy is to be a plus in helping Haiti to rebalance, 'build back better' and avoid mistakes of the past.

First, the fiasco of the 1980's 'Taiwanization' period must not be repeated. Universal *free* education and rural investment are important, and though they will not precede assembly investment, they must robustly parallel it and eventually get ahead of it. Investment in Haiti should not ignore decentralized agri-business possibilities and the economic growth and development it can bring through jobs and the infusion of cash into the Haitian economy.

Second, assembly plants cannot be concentrated largely in Port-au-Prince. If nothing else, the shattered infrastructure of the city should serve as an incentive for decentralization. Haiti has at least a dozen coastal cities that either already have a functioning, albeit usually rudimentary, infrastructure or where a port and support infrastructure can be built – perhaps at a lower costs than Port-au-Prince. Decentralization to coastal cities and towns offers Haiti and investors an opportunity to undo the damage begun fifty years ago by Papa Doc's insidious centralization in Port-au-Prince and to rebalance the prospects for economic growth and infrastructure development (including electricity) to all of Haiti.

Third, investors, owners and managers must be mindful of the fact that Haitian workers are more than plentiful cheap labor. As Secretary of State Clinton said at the April Donors' Conference, "'talent is universal; opportunity is not."[16] A key to Haiti's renaissance is to improve the opportunity environment for all of its people. Haiti's Diaspora offers bountiful evidence of what can be achieved when opportunities are twinned with talent.

Jeffry Sachs has equated factory jobs in Bangladesh with the first rung on a ladder toward greater opportunity and development.[17] In Haiti, however, the ladder for most factory workers, in view of their survival wages juxtaposed with a constantly increasing cost of living and the absence of any public social safety net, has a single rung. Haiti's opportunity environment will be improved considerably:

- If investors, owners, and mangers recognize that Haiti's workers have legitimate aspirations to improve their lives, and their honest days' work should be means for that, and;
- If investors, owners and managers follow that recognition with actions that demonstrate socially responsible investing and public-private partnerships that improve workers status and conditions, and;
- If the Haitian state has the strength and resources to become and remain a positive presence in workers lives by providing services to them and their children, particularly in education, health, and safety from gangs and other criminal elements whose activities are often financed by narcotics trafficking.

* * *

If there is a silver lining in the deep dark cloud of Haiti's recent catastrophe, it is that this offers all of us – Haitians, 'friends of Haiti' and those whose connection with Haiti may simply be as a bureaucrat or investor – an opportunity to learn from mistakes made in the relatively recent past and take steps that will rebalance that country so that it will move forward unequivocally toward less poverty and inequity, diminished social and economic exclusion, greater human dignity, a rehabilitated environment, stronger public institutions, and a national infrastructure for economic growth and investment.

Thank you.

End Notes

[1] Robert Maguire, *Haiti After the Donors' Conference: A Way Forward*, United States Institute of Peace, Washington DC, Special Report 232, September 2009.

[2] Francis Fukuyama, "Poverty, Inequality and Democracy: The Latin American Experience," *Journal of Democracy* 19, no. 4 (October 2008).

[3] Op.cit, Maguire, "*Haiti After the Donor's Conference*"

[4] Maureen Taft-Morales, "Haiti: Current Conditions and Congressional Concerns," Congressional Research Service Report for Congress, May 5, 2009.

[5] Robert Zoellick, "Securing Development" (remarks at the United States Institute of Peace conference titled *Passing the Baton,* Washington, DC January 8, 2009.

[6] Trenton Daniel, "Thousands flee capital to start anew," Miami Herald, January 23, 2010; Mitchell Landsberg, "The displaced flow into a small Haitian town," Los Angeles Times, February 1, 2010; data from MINISTAH headquarters in Hinche.

[7] Robert Maguire and Robert Muggah, "A New Deal-style corps could rebuild Haiti," Los Angeles Times, January 31, 2010.

[8] Nick Taylor, *American Made: The Enduring Legacy of the WPA*, (New York: Bantam Books, 2008), p.99.

[9] UNDP, "Fast Fact of the Week," accessed on February 2, 2010 at undp.washington@undp.org

[10] CTV.ca News Staff, "Foreign Ministers vow to be 'partners' with Haiti," accessed on January 30, 2010 at www.http://ottawa.ctv.ca/servlet/an/plocal/CTVNews/20100125/Haiticonference100125/2010

[11] Government of Haiti, "Vers un Nouveau Paradigme de Cooperation," April 2009

[12] Jacqueline Charles, "Haiti President Rene Preval quietly focuses on 'managing country," Miami Herald, February 2, 2010

[13] Hernando de Soto, *The Mystery of Capital* (New York:Basic Books, 2000).

[14] Government of Haiti, "Rapport d'évaluation des besoins après dèsastre Cyclones Fay, Gustav, Hanna et Ike," November 2008.

[15] Jim Wyss and Jacqueline Charles, "Workers flock to clothing factories as industrial park reopens," Miami Herald, January 27, 2010.

[16] Hillary Rodham Clinton, "Remarks at the Haiti Donors Conference," April 14, 2009, accessed on February 2 at http://www.state

[17] Jeffry Sachs, *The End to Poverty: Economic Possibilities for Our Times*, (New York: Penguin Press, 2005)

In: Haiti: Earthquake and Response
Editor: Rachael A. Donlon
ISBN: 978-1-61668-998-8

Chapter 11

TESTIMONY BY MARK L. SCHNEIDER, SENIOR VICE PRESIDENT, INTERNATIONAL CRISIS GROUP TO THE SENATE FOREIGN RELATIONS SUBCOMMITTEE HEARING ON INTERNATIONAL DEVELOPMENT AND FOREIGN ASSISTANCE, ECONOMIC AFFAIRS, AND INTERNATIONAL ENVIRONMENTAL PROTECTION ON HAITI RECONSTRUCTION: SMART PLANNING MOVING FORWARD, 4 FEBRUARY 2010; WASHINGTON, DC

Mr. Chairman, members of the Subcommittee, let me express my appreciation for the opportunity to offer testimony today on the immediate and long-term consequences of the earthquake in Haiti – for its people, its democracy and its neighbours.

First, I want to express my condolences to the people of Haiti for the enormous loss of human life – far more victims than in any other natural disaster in the history of this hemisphere – ever. We already know that some 150,000 people were killed, 200,000 were injured and one million more lost their homes. After all the collapsed buildings are finally removed, this earthquake may be among the three or four worst disasters ever recorded anywhere on earth in terms of loss of life and injury.

For many of us, there are faces and names we recall with a deep sense of loss. I first went to Haiti in 1978 with then Ambassador Andrew Young to raise concerns about human rights abuses under the Duvalier dictatorship. With PAHO/WHO, USAID, Peace Corps and now the Crisis Group, I have worked with Haitians desperately trying to achieve a better future for their families.

Second, let me express my deep sadness at the deaths of men and women from the UN peacekeeping mission (MINUSTAH) including its leaders Hedi Annabi, Luis da Costa, and Gerardo LeChevallier, along with Philippe Dewez from the IDB, and all of the others who were working with the government of President René Préval to improve conditions in Haiti.

Finally, let me express my own enormous pride in the generous response of citizens from this and other countries -- the volunteer doctors, nurses, NGO's and search and rescue teams, as well as the rapid and robust response from the Obama Administration, particularly USAID, State and the U.S. military, but also from Canada, Brazil, Mexico, Cuba, and others in this hemisphere and France, the EU, Spain, China and other countries outside the hemisphere.

Mr. Chairman, the important questions that you posed with respect to planning, managing and implementing Haiti's reconstruction have been the subject of much discussion in Port au Prince, at the Montreal donors preparatory session last week, at the World Economic Forum in Davos, at the UN, the EU, and the OAS. Many have looked at examples from the past – the Tsunami in Southeast Asia, Hurricane Mitch in Central America, post conflict reconstruction measures in Kosovo, El Salvador and Liberia. Each of those experiences offers lessons about relief and reconstruction, which have already helped improve the relief measures in Haiti. For just one example, Mitch taught USAID's OFDA that pre-positioning basic supplies in Florida and the Caribbean could alleviate the need for lengthy procurement procedures, and that pre-approved agreements with the Southern Command could speed transportation logistics.

All of those cases had several things in common:

The victims never felt that relief was coming as fast as they needed it;

The transition from relief to reconstruction was neither smooth nor untroubled;

Maintaining international engagement and international coordination was a constant struggle; and

The challenge of ensuring that the host government was strengthened rather than weakened was not fully met.

Given the magnitude of Haiti's destruction, the fragility of its institutions before the quake and the depth of its poverty, overcoming these challenges to effective reconstruction will pose an even more daunting challenge to Haiti and to the international community.

Mr. Chairman, the International Crisis Group has issued fifteen reports about Haiti over the past five years. The most recent, *Saving the Environment, Preventing Instability and Conflict* (April 2009) was unfortunately all too prescient in identifying the additional risks to stability and complications in urban planning, construction, and infrastructure design posed by Haiti's historical disregard for the environment and vulnerability to natural disasters.

There is a mantra now that we must help Haiti to build back better, to ensure that recovery and reconstruction leave Haiti less vulnerable to the consequences of natural disasters. That should be done. But it is also impossible to completely eliminate Haiti's vulnerability given its incredibly hazardous geologic and geographic location precariously positioned along a ghastly seismic fault line, in the annual hurricane path from Africa, and caught between the small plane and fast boat cocaine routes from Colombia and Venezuela.

However, Haiti's vulnerability also stems from its failure to overcome two centuries of bad governments, inequitable and centralized political and economic power structures in Port au Prince, and not-always-benign foreign interventions. Many point to the billions in aid that Haiti received over the last five decades and say it was all for naught, that there is no hope today.

I argue the contrary. In June, I met with several government representatives, including President Preval, and the former and current prime minister. In December, I held discussions with the late Hedi Annabi and others from the UN, IDB, WB and the representatives from

President Clinton's envoy office to assess progress and examine the challenges for 2010. There were concerns, of course, but there also was a degree of optimism:

- Reforms were taking hold within the civilian police; in fact a 2009 poll showed over 70 percent of the population approved of their performance, a far cry from the past.
- The first glimmers of judicial reform in 50 years were seen with the opening of an academy to train judges, and passage of key laws to set merit-based standards and salaries for judges and to establish a monitoring commission to vet existing judges and provide professional assessment of their performance.
- The first class of trained corrections officers had graduated and a plan to build new and restructure older jails was underway.
- The HOPE II legislation had boosted employment by close to 25,000 and recruitment by former President Clinton had brought investors to Haiti. The transition from showy pledges to actual capital investment projects underway, including on a $55 m. Royal Caribbean Cruise expansion of the Labadee resort and a new industrial park on the outskirts of Port au Prince, thanks to a $25 m. commitment from George Soros, a member of Crisis Group board of trustees.
- Haiti had a fully functioning legislature, which after risking stability by ousting a competent prime minister Michele Pierre Louis, at least demonstrated a marked readiness to act by approving the new Prime Minister Jean-Max Bellerives, his slate of ministers and their program in record time, when the same process last spring took months.
- Haiti's budget for the current fiscal year -- contrary to that of the United States -- was actually passed on time; the previous budget had not been approved until eight months into the fiscal year.
- In October, the United Nations extended its mandate for another year and Latin American nations swiftly reaffirmed its leadership, contributing some 4000 of MINUSTAH's 7000 formal military members.
- For three years, the Preval administration had met its fiscal targets, reduced inflation, and maintained a stable monetary structure. Despite the devastation caused by four consecutive storms in 2008 and the global economic crisis, Haiti was one of two countries in the region to post positive economic growth (2.4 percent) in 2009. The progress prompted the IMF and World Bank to endorse the cancellation of $1.2 billion of Haiti's multilateral debt, more than half. The earthquake not only justifies -- but truly demands – that the last half of Haiti's debt be written off.

Despite myriad problems -- some self-inflicted -- the Preval administration advanced these reforms in concert with MINUSTAH. The administration sought to engage the business community, opposing parties and civil society in developing a common vision of the future. Preval had named five ad hoc commissions, including some of his opponents and independent scholars, to identify and develop recommendations on critical issues, including the politically contentious issue of constitutional reform.

With the leadership of the current Prime Minister Jean-Max Bellerives, who was then minister of planning, the Preval administration had also partnered with local communities and multiple sectors with the support of the World Bank and UN to formulate a national consensus for poverty reduction. The result was a Haitian National Poverty Reduction

Strategy Paper, which was also endorsed by donors. After the 2008 hurricanes, the strategy was developed to include a job creation plan with a primary focus on jobs in rural agriculture, decentralized tourism, and the factory apparel industry. Donors gave their blessing to that program last April.

Those plans and strategies give Haiti a huge advantage today because they can serve as a foundation for reconstruction. In addition, some of the ideas that could not be put forward before the earthquake now can and must be considered for Haiti to transform its future.

Starting with the premise that the first phase of Haiti's reconstruction will require a decade, and the second, a generation, I offer these suggestions for five principles of successful reconstruction that could transform Haiti's political institutions and economic options.

- Forge a new Haitian Social Compact for reconstruction. A unified Haiti under its currently elected government – not any superimposed protectorate – has to be in the lead on Haiti's recovery if the effort is to be successful. Haiti's history has been defined by a small economic elite who dominated economic and political power until the 1990's, opposed tax levels needed to finance adequate state services and, in many cases, eluded their personal tax obligations as well. For Haiti's recovery to succeed, the elite must share in the sacrifice, especially since they will inevitably benefit from any success. Changing that equation will require the kind of inclusion that created the PRSP and drove a successful national advocacy campaign for Hope and HOPE II. Reconstruction has to be led by Haiti's elected government and represent all of Haiti and have the participation of the private sector. The full engagement of Haitian civil society – like the process that underpinned the PRSP – also must be generated. Communal leaders like those in NDI's Initiative committees are also potential allies in this process. Upcoming parliamentary elections have been postponed. The social compact hopefully will find a way, endorsed by all, to agree to hold the presidential, parliamentary and local elections together next November, if humanly possible, with the Constitutionally-mandated parliament remaining in office until the newly elected members take office next January.
- Build a modern Haitian state. Haitian government has always been starved for resources and its ministries have never been able to keep up with growing public needs. The reconstruction of Haiti must be aimed at transforming the country in a way that leaves a modern functioning state able to sustain public services and guarantee the rule of law. Modern communications, information technology and management systems have bypassed government ministries to some degree and denied them the capacity to actually deliver fundamental services to regional departments and municipalities. Modern data information, communications systems, and planning and evaluation capacity were all lacking in the ministries before their buildings were destroyed. Rebuilding those ministries on modern terms is essential to avoid Haiti becoming a failed state.
- Ensure economic and political decentralization. Ending centralization of virtually all economic investment in the capital is essential to reducing extreme poverty in its rural departments, and to rebuilding Port au Prince. A growing percentage of the capital's population, now estimated at close to 400,000, has returned to families in their original villages and towns, a third going to the Artibonite, originally the heart

of Haiti's rice farming. Now may be the first time that Haiti's constitutional call for decentralization can actually be attempted. If regional economic development poles can be generated around the country – for instance by implementing HOPE II in a way that encourages the construction of industrial sites in other departments with access to ports, such as Cape Haitien in the North, Port-de-Paix in the North West and St. Marc in the Artibonite – it will also help to stem the flow of migrants to Port-au-Prince. That also will give the capital a better chance for more rational reconstruction and avoid a replication of the slum communities of the past.

- Use environmental protection and disaster preparedness standards for all reconstruction projects. Haiti has gone from a country with 80% forest cover centuries ago, to about 20% in the 1940s, to two percent today. Its hillsides are mudslides waiting to happen. Every reconstruction project should be judged in part by whether it advances environmental protection, and every construction project should be judged on whether it incorporates both hurricane and earthquake resistance elements.
- Guarantee massive, coordinated assistance. The U.S. and international response must be bigger and better coordinated than ever before. The U.S. has already committed nearly $400 million to relief, and hopefully it will show leadership in formally committing to a decade-long reconstruction and development plan at the upcoming March pledging conference at the UN. While the detailed assessment of damage and reconstruction costs have yet to be completed, early estimates suggest the damage could go well beyond $10 billion. A broad group of NGOs – including the International Crisis Group – has recommended an early emergency supplemental of $3 billion as essential to Haiti's recovery. The sooner it is approved, the more likely other countries and institutions will seek a matching commitment. To put this in some perspective, in this hemisphere, the U.S. has pledged between 30-65 percent of the reconstruction aid totals following natural disasters like Hurricane Mitch or peace accords in Nicaragua and El Salvador.

For that effort to be successful, each key U.S. agency, particularly USAID, State and DoD, must designate full-time Haiti Reconstruction Coordinators. Ideally, the President should name a single Haiti Reconstruction Coordinator to serve as an overall U.S. government policy czar for Haiti reconstruction and empower him or her with the necessary authority to ensure an all-of-government response. That would ensure a greater degree of overall strategic coordination, guarantee inter-agency coherence and reduce potentially counterproductive delays.

However, the United States also must commit by example to a similar international coordinating reconstruction effort. There is already a UN peacekeeping mission on the ground. Even before the earthquake, the Secretary General's Special Representative was unable to ensure that independent UN agencies, within their competence, responded to the priorities defined by the Security Council. That needed to be changed earlier. Now it is absolutely essential. The UNSRSG also should be the interlocutor with the Government of Haiti with respect to security, rule of law and political reform and coordinate all international reconstruction assistance. In other areas, he or she should still co-chair along with the Haitian Prime Minister or the designee of the President and the Prime Minister, a technical and financial reconstruction committee, that will have the authority to review projects deemed contrary to the major objectives of the UN mandate and the goals of the Haiti reconstruction and transformation plan. Obviously the World Bank, IDB, US, EU, and others would sit on the committee with the SRSG and the Haiti government. The committee should be the mechanism of international coordination and oversee progress toward implementing the reconstruction plan and hopefully pressure each other to make good on donor pledges.

In addition, a critical Haitian government-wide procurement mechanism should be considered, in partnership with the international community, to oversee large-scale infrastructure projects proposed by Haiti for its transformation – from planning to procurement to construction to completion. Inclusion of measures of transparency and accountability in that agency will be vital not only for donor satisfaction but to avoid inevitable suspicion from Haitian constituencies as well.

Let me suggest five priority areas where many of those principles should be applied.

First, for reconstruction to succeed, both security and the rule of law are required. Reconstruction planning must incorporate a clear and critical path toward the completion of police, justice and prison reforms that were initiated before the earthquake, and deploy them across the country.

Fortunately the presence of the UN peacekeeping mission – and temporary U.S. military forces – guarantees the physical stability of the state. The past has shown us that gangs in Port-au-Prince are capable of quickly reorganizing. It appears that is what is happening now in Cite Soleil and other areas, where there are reports that criminals – many from among the 5000 prisoners who escaped the crumbled penitentiary – are resuming their criminal armed activities. The UN peacekeeping mission has been authorized for a reinforcement of 2000 more troops and 1500 more police. They will need more police, to be sure, well beyond 2011, while police stations are rebuilt and equipped and the training of new police continues. To put it in perspective – about 1000 of the 4000 police who worked in Port-au-Prince have not shown up for work or are believed to have died, although the large majority, despite their own losses in many cases, are back on the streets.

The U.S. can also respond to President Preval's pleas for help in fighting drug trafficking by boosting the interdiction capability of the Haitian coast guard and the Haiti National Police (HNP) on an on-going basis. The U.S. could also second more Haitian-American police, prosecutors and judges to the UN to assist Haiti in building its own justice infrastructure.

Second, for reconstruction to succeed, Haiti must be supported in building a nationwide system of free public elementary and secondary education – not just in Port-au-Prince but across every department. Before the quake, nearly 40% of Haiti's children were not in school. Of those in school, an estimated 80% were in private schools, most of which were unregulated, offered poor quality education, and charged exorbitant fees. The Haitian diaspora can offer unique support, particularly with teacher training. Creole-speaking former Peace Corps volunteers can play a role, and the Peace Corps already is gathering a skills-data base to link into the reconstruction effort. Supplemental funding to fund this effort should be provided.

This is also an opportunity to offer Haiti's young people a chance to participate in their country's own recovery. The concepts of AmeriCorps and the Civilian Conservation Corps should be introduced to produce jobs for the unemployed that contribute to Haiti's reconstruction.

Third, renewing Haitian agriculture may be the best way to keep the migrants from Port-au-Prince in their communities of refugee. They must have access to credit and fertilizer, assistance with marketing and perhaps even guaranteed prices for their first harvest. If that occurs, the capacity of Haitian farmers to once again be the major source of food for the population, as it was before the 1970s, would be enhanced, particularly with respect to rice. Before the 1970s, Haiti produced nearly all of its rice. Once tariffs were removed, its farmers could not compete with subsidized and large-scale rice farmers in the U.S. and they nearly disappeared, as 70 percent of Haiti's rice is now imported.

Haiti has shown that it has the potential to meet modern marketing demands with mango and coffee crops. When agriculture is linked to environmental protection with protection of watersheds, terracing and reforestation, there is a win-win outcome.

Fourth, meeting Haiti's energy requirement will be essential in any reconstruction environment and now may be the moment when an historic shift away from charcoal – as fuel for cooking and for small business energy generation – can be achieved. It would not only remove the constant threat to the nation's remaining forest cover, including in its national parks, but also enable reforestation to have some chance for success. This will require Haitian leadership with international technical and financial support in a single, unified program that subsidizes impoverished Haitians in making the transfer. This is essential along with continued reform of Haiti's electric utility, EDH.

Finally, the Haitian Social Compact should clearly engage the Haitian diaspora in the reconstruction effort. This could include providing avenues for remittances for development, with matching contributions by donors for community projects. In addition, the same concept of direct transfer of resources from a diaspora Haitian-American or Haitian-Canadian to a family member – which now surpasses official development assistance – should be used as a model for accelerating the use of conditional cash transfers to the poor, with the sole condition being that their children are immunized and attend school. Using the Brazilian, Mexican and other models, an income supplement can reach impoverished Haitian families when they need it the most.

Helping Haiti recovery from this natural disaster constitutes an obligation for every nation of this hemisphere and beyond. It is not only the right thing to do in helping neighbours, it is the only thing to do.

In: Haiti: Earthquake and Response
Editor: Rachael A. Donlon

ISBN: 978-1-61668-998-8

Chapter 12

"Haiti Relief and Recovery: Building a Better Future for Children", Charles MacCormack, President and CEO, Save the Children, Before The Senate Foreign Relations Subcommittee on International Development and Foreign Assistance, Economic Affairs and International Environmental Protection, February 4, 2010

Mr. Chairman,

Save the Children welcomes this hearing by the Senate Foreign Relations Sub-Committee on Haiti. It raises important issues that deserve full discussion because the choices will affect lives and the future of a nation.

I have traveled to Port-au-Prince twice since the January 12 earthquake and saw firsthand its catastrophic impact. The devastation of the capital, with its highly centralized government infrastructure and institutions, is reverberating across the entire country. No part of Haiti is unaffected, as thousands of displaced families and children leave the earthquake zone, and as insufficient government services and infrastructure falter or collapse.

Children are always among the most vulnerable during emergencies. Sadly, the damage wrought by this earthquake only compounds the challenges that Haiti's children and families already faced each day. Their needs are enormous: medical care, food and water, shelter, basic supplies, protection and education. The threat of disease and illness is constant and cases of tetanus and suspected cases of measles have already been reported. Children are separated from their families, and are at risk for abuse and exploitation, as well as psychosocial distress. While the Haitian people are extremely resilient and have exhibited much patience, their challenges are daunting.

The earthquake caused severe damage, but also created the need for visionary thinking towards recovery and reconstruction so Haiti can start a new chapter and proceed on a new path forward. Save the Children hopes to work with Haitians to turn that page, based on our experience and expertise from over three decades in the country working in a complex environment characterized by frequent humanitarian crises. We have launched in Haiti one our largest international responses ever. Save the Children is now engaged both in an immediate response to a disaster of unprecedented scale as well as in development of a bold and ambitious vision that is both long-term and comprehensive, incorporating principles of supporting Haitians "building back better for children" at every step.

Our goal is to provide emergency assistance to save lives, alleviate suffering and support the recovery of 800,000 people including 470,000 children. This effort is focused on the entire country. There are no unaffected areas, so we all need to develop solutions both for those in the earthquake zone as well as those elsewhere, such as enhancing secondary city habitability and improving rural agricultural zones. We are working together with the government of Haiti and other international and local organizations to assess and respond to the needs of children without parental care and to identify, register and reunite separated children with their families. We have also accepted the United Nation's request to help coordinate efforts to reunite separated children and co-lead the international response on education. In addition, we are delivering medical supplies to hospitals, distributing food, opening mobile health clinics and creating child-friendly spaces for children.

Our vision is a Haiti where all children realize their rights every day to a basic education, a healthy life, freedom from abuse and benefit from the support of families who recognize the fundamental needs of their children. In order to make lasting, positive change in the lives of children, Save the Children calls on the U.S. Government to sustain robust support for meeting Haiti's immediate and long-term development needs in cooperation with the Haitian government and a wide variety of partners, including local civil society and children themselves.

During all stages from early recovery through reconstruction, the international community and Haitian authorities must demonstrate a serious commitment to disaster risk reduction in all spheres of activity. Hurricane season is just around the corner and we know these kinds of investments up front can avoid much greater costs incurred later through humanitarian responses.

Haiti's Rebuilding Will Require Substantial Investment

The international community must fulfill the United Nations flash appeal for $576 million, and then sustain significant investments for the next ten years. We need to ensure the international community stays the course and that, unlike in past humanitarian crises, attention does not erode as other challenges arise. The government of President Preval has shown for the past two years that it is a government that the international community can work with – we must make our commitments long-term, predictable and transparent.

In 2008 Congress passed HOPE II (the Haitian Hemispheric Opportunity through Partnership Encouragement (HOPE) Act), which extended trade preferences to Haiti for ten years. As others have recommended, Congress should consider broadening these preferences to include even more of Haiti's exports. Finally, instead of issuing new loans to the Haitian government, post-earthquake assistance should be in the form of grants. Outstanding bilateral and multilateral debt of nearly one billion dollars should be cancelled. Future funds must go to rebuild Haiti and ensure its children have a future to look forward to.

Haiti's Rebuilding Will Take Time

The recent Ministerial planning conference in Montreal should the first step of a ten-year commitment by the international community to walk with Haitians toward a new future and to meet their immediate and long-term development needs in cooperation with a wide variety of partners, including local civil society and the government of Haiti. Save the Children has been in Haiti for over 30 years. Today when we start working with a community, we want to make a commitment to them – to work with them for ten years and build their capacity during that time to take over from us. We work with the government, with local nongovernmental organizations (NGOs) and with the community. We can only make such a commitment with support from our private donors because few institutional donors have a ten-year timeline. But Haiti needs such a long term commitment from all of us.

Enhanced Coordination Is Also Necessary

We all need to do better at coordinating our efforts, especially given the thinly stretched capacity of the Haitian government. Even before the earthquake, coordination among the major donors, NGOs and others was more a case of information sharing than thoughtful division of labor. With more to accomplish and more actors engaged, the importance of strategic coordination cannot be understated.

To do our part, Save the Children is playing a leading role in the UN cluster system, which seeks to coordinate the international humanitarian response by sector. Previous experience suggests, however, that in working alongside and through the UN system, the U.S. should empower one concerned agency to oversee the overall response. President Obama wisely empowered Dr. Rajiv Shah, the new Administrator of the U.S. Agency for International Development (USAID), to oversee the U.S. relief response in Haiti. USAID should be empowered as well to lead the U.S. development effort, so that the dozens of U.S. agencies engaged with Haiti respond in a coordinated and cohesive way to support Haiti's long-term development needs.

Some are questioning whether longer term aid will be well spent given the political instability and corruption that were wide spread prior to the earthquake. The answer, of course, depends on the Haitian people, but the U.S. government can do its part to move in the right direction. As the largest donor to Haiti in the past, the effectiveness of the U.S. government programs impacts the entire development enterprise in Haiti. Just a few donors, the U.S., Canada, France, the EC and the World Bank, contributed over 90 percent of the official development assistance in 2006 and 2007 with the U.S. and Canada alone providing almost half. Better and more strategic coordination amongst the international community is both possible and necessary.

Working with the Haitian government and civil society, the U.S. can make significant progress towards more effective reconstruction and development. The U.S., with NGO partners and other donors, should intensify its commitment to building the capacity and systems of the Haitian government and Haitian civil society to lead and manage their own development.

Rebuilding infrastructure is important; building institutions, as former USAID Administrator Natsios argued in a recent essay, is of even greater importance. In Haiti, Save the Children will continue to work with Haitian institutions and build Haitian systems. For instance, prior to the earthquake we were supporting 250 schools in Haiti – government, religious and private schools. Schools supported by the government are apparently the strongest, but the government did not, and certainly now does not, have the capacity to spend even the few funds that the international community provides to it. As schools reopen, we will continue to support school health and nutrition, and teacher training at these schools, and work with these schools to go through the government certification process which ensures that the schools are aligned with government protocols. The U.S. and international partners must strengthen both government institutions to oversee and provide education, and the private and non-profit Haitian institutions to link with this system.

Drawing on Save the Children's previous experience, including lessons learned by the 2004-2005 Tsunami response in southeast Asia, we know that Haiti's effective long-term development will require putting Haitians at the center of their own development and recognizing the critical role of women and youth in the decision making process. In this regard, we believe the Administration and Congress should consider the following:

- Invest in participatory initiatives that engage women and children in the decision-making, implementation and monitoring of reconstruction and development initiatives;
- Focus on rebuilding Haitian institutions and systems. Infrastructure matters a great deal, but promoting human development by equipping Haitians to deliver quality education, healthcare and other services themselves matters even more.
- Explore models for attracting back the Haitian diaspora into the government at reasonable salaries, as Liberia and Afghanistan have done with some success.
- View the Haitian government's recently completed development plan as the primary plan for Haiti's reconstruction and development.
- Strengthen USAID with more staff to properly plan, engage, leverage with others and monitor our assistance. Currently there too few staff and they spending 30 percent of their time on reports and planning documents for Washington DC.

Business as usual is not enough for Haiti. The international community, the U.S. and the NGOs must sharply expand our focus on human development, both skills and institutions.

Governments, NGOs, foundations and other stakeholders must help build the capacity of Haiti's civil society, private sector and national government from the very local to the national level to enable Haitians to lead their country into a brighter future.

Thank you.

INDEX

#

A

B

C

E

H

I

J

K

L

M

N

O

P

Q

R